2nd International Conference on Advanced Nanomaterials and Nanodevices (ICANN 2019)

IOP Conference Series: Materials Science and Engineering
Volume 813

Shanghai, China
23 - 25 October 2019

ISBN: 978-1-7138-2675-0
ISSN: 1757-8981

Printed from e-media with permission by:

Curran Associates, Inc.
57 Morehouse Lane
Red Hook, NY 12571

Some format issues inherent in the e-media version may also appear in this print version.

Institute of Physics
Dirac House, Temple Back
Bristol BS1 6BE UK

Phone: 44 1 17 929 7481
Fax: 44 1 17 920 0979

techtracking@iop.org

Additional copies of this publication are available from:

Curran Associates, Inc.
57 Morehouse Lane
Red Hook, NY 12571 USA
Phone: 845-758-0400
Fax: 845-758-2633
Email: curran@proceedings.com
Web: www.proceedings.com

TABLE OF CONTENTS

IOP Conf. Series: Materials Science and Engineering **813** (2020) 011001 doi:10.1088/1757-899X/813/1/011001

Preface

On behalf of the organizing committee, we are pleased to present you with this volume. It contains the papers accepted for presentation in the conference of 2019 2nd International Conference on Advanced Nanomaterials and Nanodevices (ICANN 2019) which was successfully held in Shanghai University, China during October 23-25, 2019.

The topics are covered in the ICANN include Computational Nanotechnology, Graphene Managementof Nanotechnology, Nano-measurement Devices, Nano-porous Materials, Nanocatalysis, Nanocomposites and so on. The conference provides a wide platform to the conference participants to share their research achievements and communicate with each other. And the distinguished keynote speakers of the conference presented their excellent academic research at the conference which has promoted the academic atmosphere. Also, reviewing process of the ICANN was a challenge process that relies on goodwill of those people involved in this field. We invited more than many researchers from related fields to review papers for the publication in the IOP conference proceedings.

Meanwhile, we would like to thank all the authors of submitted papers for choosing ICANN 2019 and thank you all the reviewers for their time and effort in reviewing articles. Finally, it is a great honor to thank you all of behind staff who made their constant effort to make sure the conference process can go smoothly and the proceeding team who have dedicated their constant support and countless time to bring these scratches into this volume.

Conference Chair
Prof. Yonghui Deng
Fudan University, China

Conference Co-Chairs

Prof. Yonghui Deng, Fudan University, China
Prof. Jiaqiang Xu, Shanghai University, China
Prof. Gong Hao, National University of Singapore, Singapore
Assoc. Prof. Ahmed Abdelfattah Elzatahry, Qatar University, Qatar

Program Chairs

Prof. Chengzhong (Michael) Yu, The University of Queensland, Australia
Assoc. Prof. Xiaowei Cheng, Fudan University, China

Technical Committees

Prof. Hikmat S. Hilal, An-Najah National University, Palestine
Prof. Mário Guerreiro Silva Ferreira, University of Aveiro, Portugal
Prof. Nikolai A. Sobolev, University of Aveiro, Portugal
Prof. Vincent Teng, Swansea University, UK
Prof. Toshio Kamiya, Tokyo Institute of Technology, Japan
Prof. Jianzhong Lin, Zhejiang University, China
Prof. Yongfeng Shen, Northeastern University, China
Prof. Bin Zhang, Northeastern University, China
Prof. Wenli Pei, Northeastern University, China
Prof. Jitladda T.Sakdapipanich, Mahidol University, Thailand
Assoc. Prof. Sotirios Baskoutas, University of Patras , Greece
Assoc. Prof. HE Weidong, University of Science and Technology of China, China
Assoc. Prof. Xiaohui Ning, Xi'an Jiaotong University, China
Assoc. Prof. Vuthichai Ervithayasuporn, Mahidol University, Thailand
Assoc. Prof. Waail Mahmod Lafta, Griffith University, Australia
Asst. Prof. Mehdi Khodaei K. N. Toosi University of Technology, Iran
Dr. Shengyu Jin， Purdue University, West Lafayette, USA

First Name	Last Name	Institution	Email Address	
Hamzah	Almadani	Ningbo University, China	almdanih@yahoo.c	A005
Zhou	Yuan	City University of Hong	zhouyuan1117@12	A001
Zhao	Yan	Beihang University,	zy_buaa@buaa.edu	A002
Meng	Zhang	Beijing Institute of	zhangmeng123@bu	A004
Bing	Jia	Science and Technology	jbrobin@sina.com	A004
DU	Liang	China Special Equipment	duliang.sme@163.c	A006
Jie	Zhang	Air Force Engineering	afeu_zhangjie@163	A007

IOP Conf. Series: Materials Science and Engineering **813** (2020) 011004 doi:10.1088/1757-899X/813/1/011004

Peer review statement

All papers published in this volume of *IOP Conference Series: Materials Science and Engineering* have been peer reviewed through processes administered by the proceedings Editors. Reviews were conducted by expert referees to the professional and scientific standards expected of a proceedings journal published by IOP Publishing.

IOP Conf. Series: Materials Science and Engineering **813** (2020) 012001 doi:10.1088/1757-899X/813/1/012001

A Multifunctional Ankle Foot Orthosis Utilizing a Magnetorheological Actuator

Y Zhou and L Liu

Department of Biomedical Engineering, City University of Hong Kong, Hong Kong

luliu45@cityu.edu.hk

Abstract. The intelligent material, so-called magnetorheological (MR) fluid, is used in the design of an ankle foot orthosis. This paper presents the configuration of a multifunctional ankle foot orthosis that is equipped with a novel MR actuator. At first, the structure of the orthosis is introduced. Then, according to the mounting position of the actuator, the MR actuator is conceived by designing the structure and magnetic circuit. The actuator has three working modes, assistant, resistant and free modes. Finally, the manual power consumptions of two conditions, gait with and without the orthosis, are investigated; electricity consumptions of two cases, orthosis with and without the MR actuator, are compared.

1. Introduction

Given the global issue of aging, in recent decades, orthotic treatment has become a popular method to rectify abnormal gait styles, due to the portability, relative low cost and ease of use. In particular, the ankle foot orthosis (AFO) is one of the most commonly used orthotic devices.

In the early stage, AFOs were passive and a certain AFO was only suitable to a certain subject. In recent years, owing to the advance in smart materials and intelligent control technologies, active and multifunctional AFOs have been developed. Generally, an AFO is equipped with an actuator, which transmits the controllable energy from the power source to the ankle joint or dissipates energy of the shank. These actuators include pneumatic rotary actuator [1], hydraulic cylinder [2], series elastic actuator (SEA) [3], shape memory alloy [4] and electrorheological (ER) fluid [5]. However, it is found that pneumatic and hydraulic actuators have the problems of portability; ER fluids should be provided with high voltage.

Some researchers shifted the attention to development of orthosis that is equipped with MR fluid. MR fluid has the inherent advantage of quick response, low energy consumption and long life period. Kikuchi et al. [6] designed an AFO with an MR brake; this brake generates resistant torque of 10 Nm and can prevent foot drop effectively. Chen and Liao [7] presented an MR actuator that can transmit unidirectional assistant or resistant torque of 30 Nm. The switch between assistant and resistant torque modes is realized by setting a locking mechanism manually. Ma [8] conducted optimal design of an MR brake, which is used to generate resistant torque at the knee joint.

During a normal gait cycle, both changeable resistant and assistant torques are required at the ankle joint. To the best of our knowledge, few existing MR-based orthotic devices are able to transmit both changeable assistant and resistant torques. Therefore, in this study, a multifunctional MR actuator that can transmit changeable assistant and resistant torques is investigated. This paper is organized as follows. Firstly, the configuration of the ankle foot orthosis is presented. The following section is the

IOP Conf. Series: Materials Science and Engineering **813** (2020) 012001 doi:10.1088/1757-899X/813/1/012001

design of the MR actuator, both in structure and magnetic circuit. Last section investigates power consumption.

2. Configuration of Orthosis

2.1. Structure

The configuration of ankle foot orthosis is illustrated in figure 1. It is mainly composed of motor, gear box, bevel gear pair, MR actuator, main frame, fixing belt and insole. Force sensitive resistors are placed on the bottom plate; they are used to measure the ground reaction force and further detect the gait events. Furthermore, fixing belt is designed at the main frame and the bottom plate to attach the orthosis to the shank and the foot firmly.

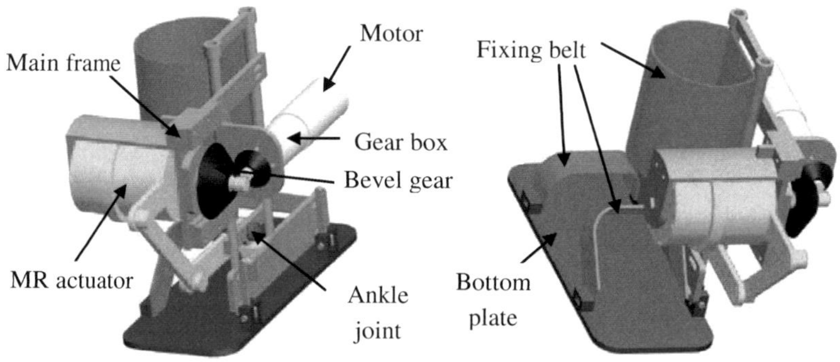

Figure 1. Configuration of the orthosis.

2.2. Function

The orthosis has three functions. Firstly, it can provide the changeable assistant torque. Subjects with weak muscle or motor control disorder can obtain the assistive torque and achieve the normal gait. Secondly, it generates the changeable resistant torque, which helps to overcome symptoms such as foot drop and get normal gait. Thirdly, it works in free mode. Subjects can walk almost freely and without the effect of the orthosis.

3. MR actuator

3.1. Structure

Figure 2 is the assembly of the actuator. It can be divided into three parts: part 1, 2 and 3; they are in the color of blue, green and yellow respectively. Part 1 is connected with the motor; while part 3 is fixed on the main frame of the orthosis. Part 1 and 2 can rotate independently relative to the main frame of the orthosis.

Each gap between inner polar plate and exterior polar plate is filled with MR fluids. The MR fluid is an intelligent material. Under the effect of magnetic field, its rheologic property will change steeply. Normally, the shear stress yield of MR fluids can be illustrated by Bingham Model:

$$\begin{aligned} \tau &= \tau_0(H)\operatorname{sgn}(\dot{\gamma}) + \eta\dot{\gamma} \quad & |\tau| > |\tau_0| \\ \dot{\gamma} &= 0 & |\tau| < |\tau_0| \end{aligned} \tag{1}$$

where τ is shear stress yield; $\dot{\gamma}$ is shear strain rate; η is viscosity; τ_0 is mangneto-shear stress yield; H is magnetic field intensity.

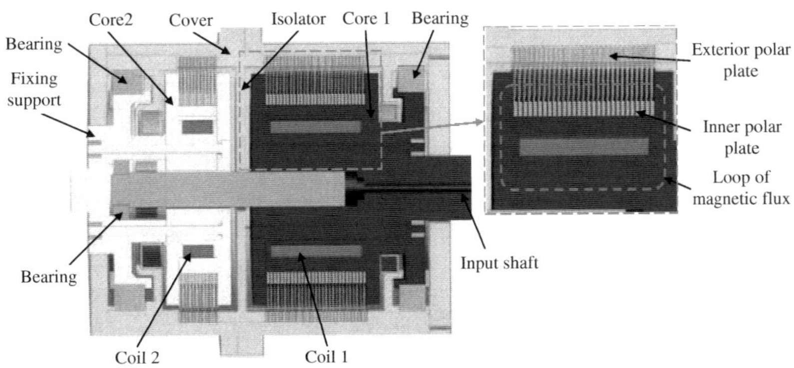

Figure 2. Magnetorheological actuator.

Based on the configuration of the actuator, the MR fluids is in shear working mode, shear strain rate $\dot{\gamma}$ can be expressed as

$$\dot{\gamma} = \frac{\omega r}{w} \tag{2}$$

where ω is the relative rotating speed of approaching polar plates; r is the radius; w is the gap between approaching external and internal polar plates.

3.2. Working modes

The MR actuator has three working modes: assistant, resistant and free modes. The details of each working modes are described in table 1:

Table 1. Working modes of the actuator.

Working mode	Output torque	Coil 1	Coil 2
Assistant	Assistant torque	Switch on	Switch off
Resistant	Resistant torque	Switch off	Switch on
Free	Zero-field torque	Switch off	Switch off

In assistant mode, coil 1 is switched on and the MR fluids between part 1 and 2 undergoes rheology. Accordingly, the assistant torque generated by the actuator under post-yield state is:

$$T_{ass} = 2N_{p1}\int_{r_1}^{r_2} 2\tau_2 \pi r^2 dr - 2N_{p2}\int_{r_1}^{r_2} 2\tau_1 \pi r^2 dr \tag{3}$$

where N_{p1} and N_{p2} are the number of inner polar plate 1 and 2 respectively; r_2 is the external radius of inner polar plate 2; r_1 is internal radius of exterior polar plate 2; τ_1 is the zero-field shear stress of fluids between part 2 and 3, that is $\tau_1 = \eta\dot{\gamma}_1$;$\tau_2$ is the shear stress of fluids between part 1 and 2 of the actuator, that is $\tau_2 = \tau_{02} + \eta\dot{\gamma}_2$, where $\dot{\gamma}_1$ and $\dot{\gamma}_2$ are shear strain rates of fluids between part 1 and 2, and part 2 and 3 of the actuator respectively.

Using equation (1)-(3), assistant torque becomes

$$T_{ass} = \frac{4N_{p1}\pi\tau_{02}(r_2^3 - r_1^3)}{3} + \frac{\pi\eta(N_{p1}\omega_1 - N_{p2}\omega_2)(r_2^4 - r_1^4)}{w} \tag{4}$$

where ω_1 and ω_2 are relative rotating speeds between part 1 and 2, and part 2 and 3 respectively.

Similarly, resistant and free-mode torques can be calculated as follows:

$$T_{res} = \frac{4N_{p2}\pi\tau_{01}(r_2^3 - r_1^3)}{3} + \frac{\pi\eta(N_{p1}\omega_1 + N_{p2}\omega_2)(r_2^4 - r_1^4)}{w} \quad (5)$$

$$T_{free} = \frac{\pi\eta(N_{p1}\omega_1 + N_{p2}\omega_2)(r_2^4 - r_1^4)}{w} \quad (6)$$

where τ_{01} is mangneto-shear stress yield of fluids between part 2 and 3.

3.3. Magnetic Circuit

3.3.1. Materials of components. The electrical pure iron is chosen as the material of iron core due to its high relative magnetic permeability. Similarly, silicon-steel sheet is used to produce inner and exterior polar plates. Since magnetic flux leakage of the circuit should be prevented, cover is fabricated of aluminum because it has extremely high magnetic resistance.

3.3.2. Simulations. Simulation of magnetic circuit is conducted using ANSYS. Firstly, a geometrical mode is set up. Then, the geometrical model is meshed using element PLANE53 and each coil is applied with current of 1.5 A. The following graphs present some simulation results.

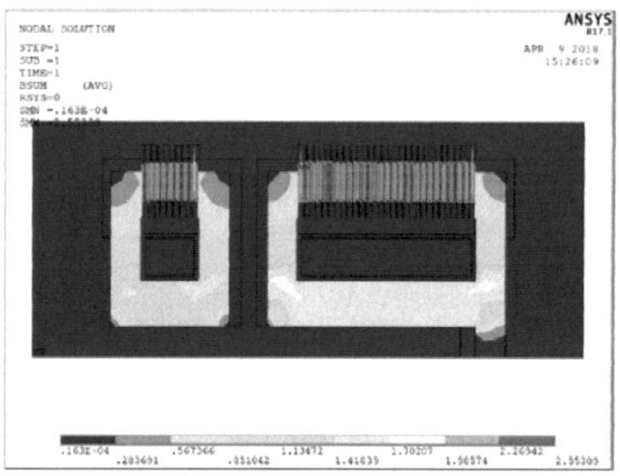

Figure 3. Magnetic flux density.

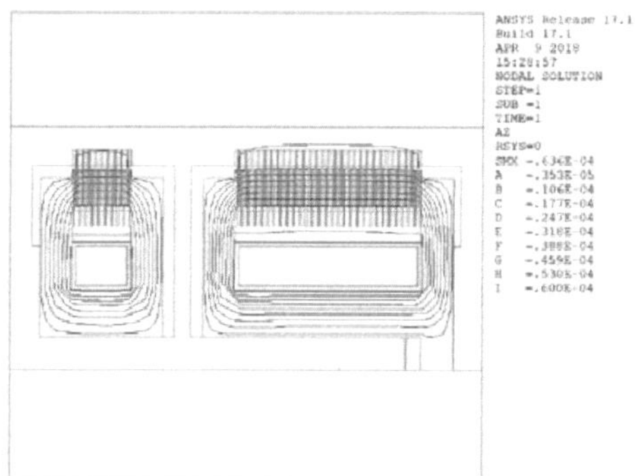

Figure 4. Magnetic flux lines.

According to figure 3, the maximum magnetic flux density in most regions is less than 2 T and the average magnetic flux densities in damping gap (areas between inner and exterior polar plates) of part 1 and 3 are 0.408 T and 0.396 T respectively, which are in the normal scope. In figure 4, there is almost no magnetic flux line that traverses across part 1 and part 3. It means the actuator's assistant or resistant working mode can be realized independently.

4. Power Consumption

4.1. Save of manual power

Relying on simulations, this part explores the comparison of manual power consumption in two conditions, walking with and without orthosis. Suppose gait parameters, such as walking speed and joint velocities, are identical in two conditions, the difference between energy consumption of orthosis-based gait $W_{\text{man,orth}}$ and nature gait $W_{\text{man,natu}}$ can be described as follows:

$$\Delta W = W_{\text{man,orth}} - W_{\text{man,natu}} = W_{\text{orth,tq}} - W_{\text{orth,wt}} \quad (7)$$

where $W_{\text{orth,tq}}$ is the work of the input torque from the orthosis; $W_{\text{orth,wt}}$ is the work of subject to overcome the weight of the orthosis. They can be calculated as:

$$W_{\text{orth,tq}} = \int_T P_{\text{orth,tq}}(t)dt = \int_T T_{\text{actu}}(t)\omega_{\text{ak}}(t)dt. \tag{8}$$

$$W_{\text{orth,wt}} \approx \int_T P_{\text{orth,wt}}(t)dt = \int_T T_{\text{orth,ak,st}}(t)\omega_{\text{sh,st}}(t)dt + \int_T T_{\text{orth,kn,sw}}(t)\omega_{\text{sh,sw}}(t)dt + \int_T T_{\text{orth,hi,sw}}(t)\omega_{\text{th,sw}}(t)dt. \tag{9}$$

where $P_{\text{orth,tq}}(t)$ and $T_{\text{actu}}(t)$ are power and assistant torque provided by the orthosis respectively; $\omega_{\text{ak}}(t)$ is normal ankle rotation speed; $P_{\text{orth,wt}}(t)$ is power caused by the weight of orthosis; $T_{\text{orth,ak,st}}(t)$ and $\omega_{\text{sh,st}}(t)$ are torque of orthosis at the ankle joint and normal shank rotation speed at stance stage; $T_{\text{orth,kn,sw}}(t)$ and $T_{\text{orth,hi,sw}}(t)$ are torques of orthosis at knee and hip joints at swing stage respectively; $\omega_{\text{sh,sw}}(t)$ and $\omega_{\text{th,sw}}(t)$ are normal shank and thigh rotation speeds at swing stage respectively.

Some parameters are chosen as follows: shank length 0.396 m, weight of orthosis 2.5 kg and $T_{\text{actu}}(t)$ is set as half of the normal ankle torque; figure 5 shows the results, $P_{\text{orth,tq}}$ and $P_{\text{orth,wt}}$. Further, during a normal gait cycle, $W_{\text{orth,tq}}$ and $W_{\text{orth,wt}}$ are 15.3 J and 5.36 J respectively; it means that positive function of the orthosis outweighs its negative function.

4.2. Electricity consumption

Electricity consumption is mainly caused by motor and MR actuator. Here, two cases of electricity consumption are considered. Specifically, case 1 is the orthosis that is equipped with a motor and MR actuator; case 2 is the orthosis that is equipped with a motor only.

Further, the assistant torque is set as half of the normal torque when the normal torque is not larger than 40 Nm; otherwise, the assistant torque is defined as 20 Nm. Figure 6 presents the simulation results. It shows that the electricity power of case 1 is slightly larger than that of case 2. Therefore, addition of MR actuator will not exert burden of energy consumption.

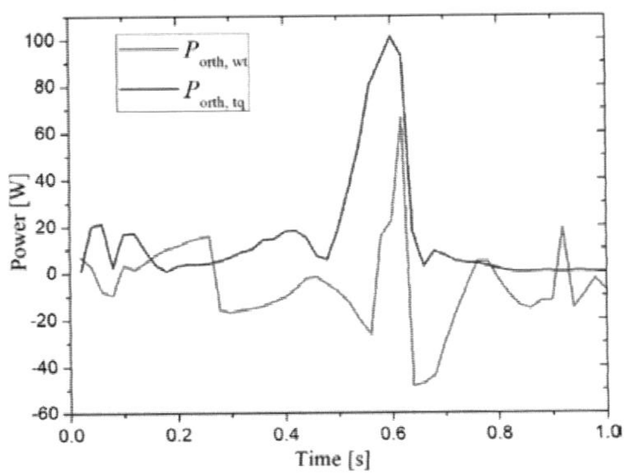

Figure 5. Powers of output torque and weight of the orthosis.

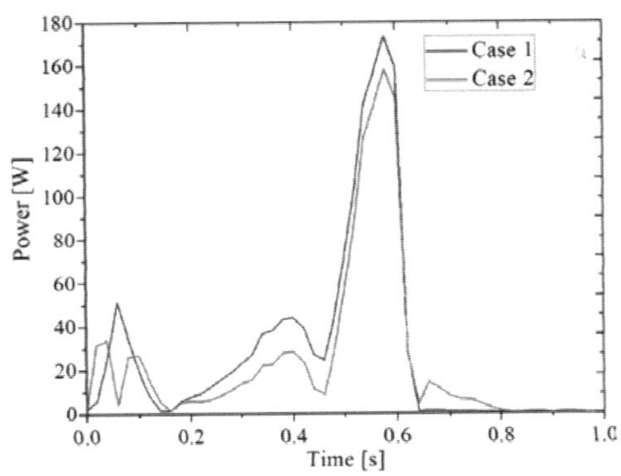

Figure 6. Electricity power consumption of two cases.

5. Conclusions

In this work, the configuration of a multifunctional ankle foot orthosis was presented. The key component of the orthosis, a novel MR actuator, was developed by designing the structure and magnetic circuit. The actuator can transmit both bidirectional changeable assistant and resistant torques. Further, the magnetic circuit of the actuator was simulated using ANSYS.

IOP Conf. Series: Materials Science and Engineering **813** (2020) 012001 doi:10.1088/1757-899X/813/1/012001

Then, advantages of the MR-based orthosis were analyzed. Specifically, the manual power consumptions of two conditions, gait with and without the orthosis, were studied; electricity consumptions of two cases, orthosis with and without the MR actuator, were compared. Results indicated that the orthosis can relieve the manual power of subjects; difference of electricity consumptions of two cases is not significant. In the future, design factors including the weight and orientation of the ankle joint's axis will be considered to improve the feasibility of the device.

6. References

[1] Wang Z and Hsiao-Wecksler E T 2016 *J. Med. Devices.* **10** 030963

[2] Neubauer B and Durfee W 2016 *ASME 2016 Dynamic Systems and Control Conference.* American Society of Mechanical Engineers Digital Collection

[3] Erdogan A, Celebi B, Satici A C, and Patoglu V 2017 *Auton Robot.* **41** 743-758

[4] Deberg L, Taheri Andani M, Hosseinipour M and Elahinia M 2014: *Smart Materials Research.* **26** 639-651

[5] Nikitczuk J, Weinberg B and Mavroidis C 2007 *Smart Mater. Struct.* **16** 418

[6] Kikuchi T, Tanida S, Otsuki K, Yasuda T and Furusho J 2010 *2010 IEEE International Conference on Robotics and Automation* p 2209-2214

[7] Chen J Z and Liao W H 2009 *2008 IEEE International Conference on Robotics and Biomimetics* p 512-517

[8] Ma H 2015 *PhD thesis*, Chinese University of Hong Kong

The Simulated Distortion Spectrum on Aluminum Alloy Structure with the Fatigue Crack Propagation

Yan Zhao[1,a] ,Weifang Zhang[2,b], Bo Jin[3,*,c], Meng Zhang[4], Wei Dai[5]

[1,3] School of Energy and Power Engineering, Collaborative Innovation Centre for Advanced Aero-Engine, Beihang University, Beijing, China.
[2,4,5] School of Reliability and Systems Engineering, Beihang University, Beijing, China.

[a]zy_buaa@buaa.edu.cn, [b]08590@buaa.edu.cn, [c]by1504121@buaa.edu.cn

Abstract. Fiber Bragg Grating (FBG) is expected to be a means of real-time monitoring of crack damage at the edge of the hole due to the sensitive perception of strain. In this paper, an optical fiber fatigue monitoring test platform is built for real-time monitoring of hole edge crack damage. Experimental results found that the grating could be sensed the strain variation with the crack propagation, caused by crack singularity, and the spectrum is highly correlated with the crack length. Full width at half maximum (FWHM) is a type of damage feature extracted from the simulated distortion spectrum, which has a high correlation with crack propagation. Meanwhile, the signal processing technique is used for the FWHM extraction, and the spectrum simulation, combined transfer matrix method and the finite element method, is investigated in this paper. The results show that the FWHM broadened obviously with the crack length, and the FWHM can be used as an effective index for evaluating fatigue crack growth.

1. Introduction

Fiber Bragg grating sensors in structural health monitoring (SHM) system has shown giant development space [1]. The FBG sensor has the characteristics of multiplexing and small size, etc. And it has been widely used in strain monitoring of nuclear power stations and large dams [2-4]. Simultaneously, the change of the physical properties of the composite structure is studied deeply because of the non-uniform strain distribution along the grating caused by the crack or layer damage in the composite structure [5]. However, the application of FBG sensor in crack damage detection of aluminum structure is a challenge all the time.

FBG sensors have been deliberately placed in non-uniform strain fields by some researchers [6], and the FBG sensors had been used for the composite health monitoring. In order to locate micro-cracks and delamination in the composite structure, the composite structure is embedded into bridging fibers [7]. Meanwhile, the spectrum width demonstrated a high correspondence with the crack density. Rajabzadeh [8] uses spectral bandwidth to measure the damage accumulation degree of composites quantitatively. However, few experimental researches have studied the use of FBG sensors adhered on the aluminum plates for the crack propagation. Jin [9] researched the crack propagation on the metal plate corresponding to the experimental spectrum change of the FBG sensor, and found that the spectrum is completely distorted, such as the emergence of multi-peaks in the distorted spectrum.

In this paper, we extracted FWHM damage characteristic by the peak seek algorithm, and the deeply relationship between the FWHM with the crack length was further researched.

IOP Conf. Series: Materials Science and Engineering **813** (2020) 012002 doi:10.1088/1757-899X/813/1/012002

First, the spectral reconstruction algorithm method was given. Because of the mechanical and optical inhomogeneity in the grating, the non-uniform strain can not be reconstructed directly under uniform strain condition.

Cięszczyk [10] given the mathematical optimization model constrained by the main wavelength and spectral region, and estimated the shape of the strip. But these similarly technologies for the optical parameter extraction were usually costly.

Under non-uniform strain conditions, Jenkins and Joyce indicate that the transfer matrix method is a good way to predict spectral [11].

The paper shows that spectral reconstruction is realized, and TMM which is a numerical method is combined with the finite element model (FEM). After that, according to the spectral analysis, the corresponding damage feature extraction methods are put forward.

2. Finite Element Method Model

In order to analyze the crack growth problem, it is combined with structural health monitoring technique. The FEM of aluminum alloy specimen was founded by using commercial software ANSYS. Moreover, the strain got under different crack lengths was monitored by finite element model, and the strain value along the grating is obtained. They are shown in Figure. 1.

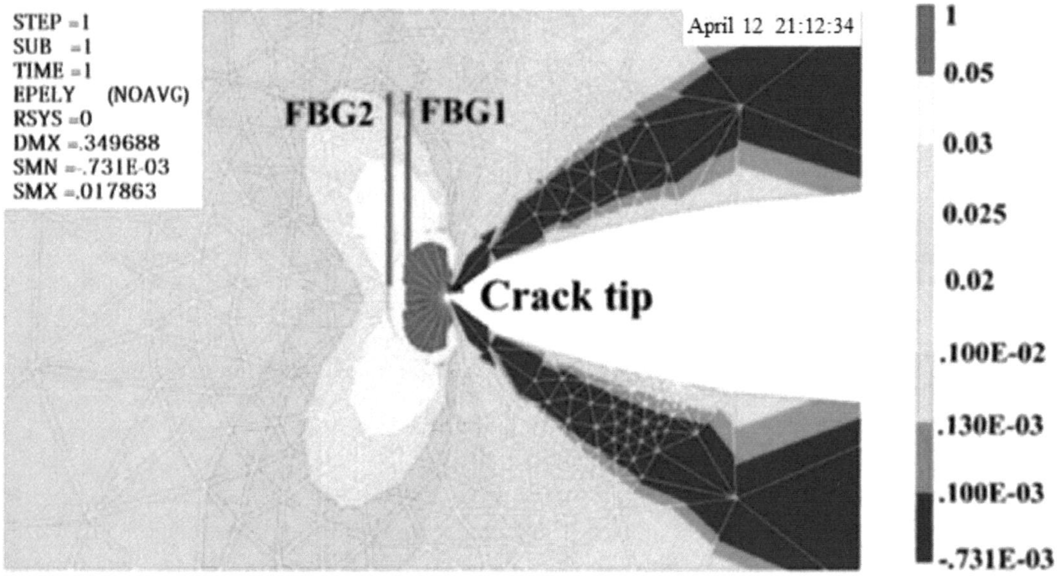

Figure 1. Strain distribution along the grating versus crack length

3. The spectral simulation

The results show that the strain pattern is distributed near the crack tip and is completely non-uniform. In our previous studies, when the tip of the crack passes through the grating, the spectrum will be distorted. In addition, the relationship between the whole length of grating and the spectrum of inhomogeneous strain images during fatigue crack growth is also studied. By combining the finite element method with the TMM method, the spectrum simulation method is proposed.

The strain distribution along the grating is obtained by finite element analysis using commercial software ANSYS. In the numerical spectrum reconstruction method TMM, the calculated strain pattern is introduced and the FBG spectrum intensity is obtained. Figure. 2 shows the structure of the spectral simulation algorithm.

IOP Conf. Series: Materials Science and Engineering **813** (2020) 012002 doi:10.1088/1757-899X/813/1/012002

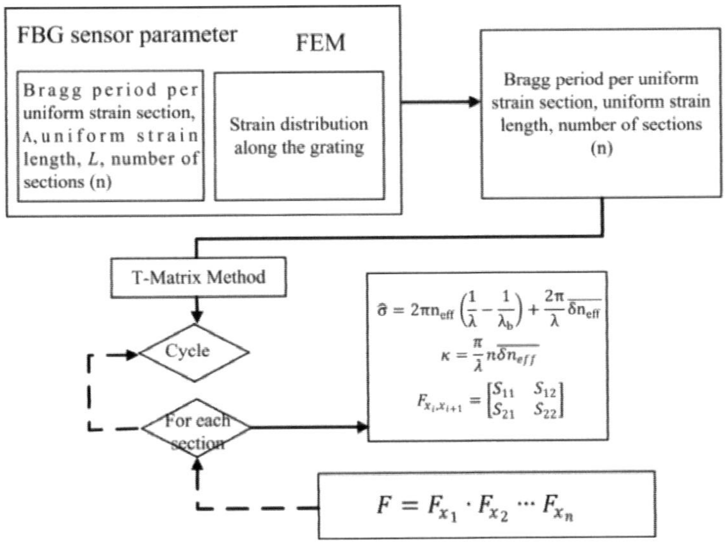

Figure 2. FBG spectrum simulation algorithm structure

The spectral changes corresponding to crack propagation are illustrated by spectrum simulation method. Firstly, FEM is carried out to better understand the strain distribution in the process of crack propagation. In addition, the strain value along the grating is synthesized into an equation which is in good agreement with the crack growth. Figure. 3 shows the spectrum distortion in various strain distribution.

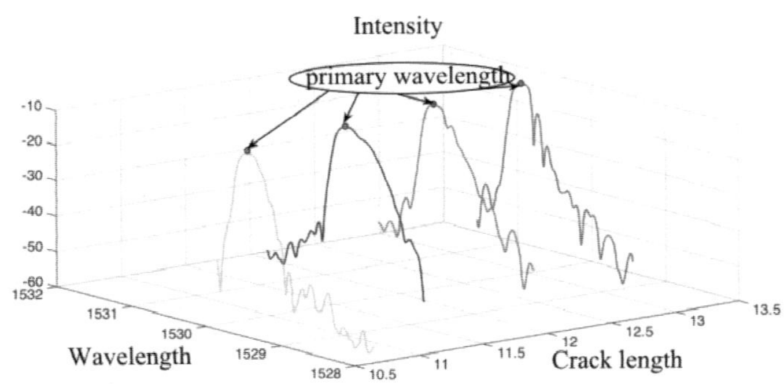

(a) The spectra under quadratic strain pattern

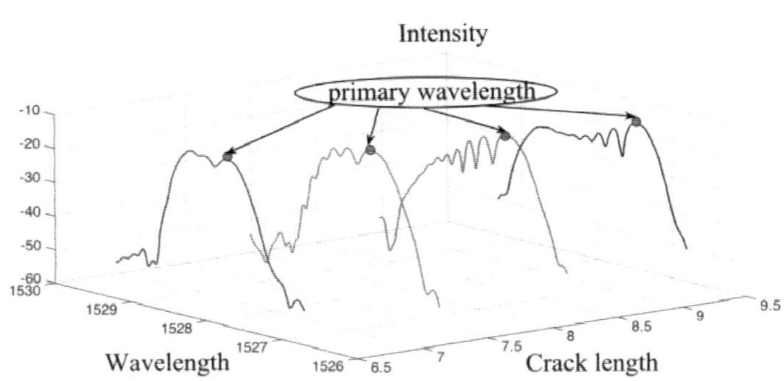

(b) The spectra under cubic strain pattern

Figure 3. FBG spectrum under difference strain pattern

4. FWHM extraction algorithm

It is apparently that the strain distributed is completely non-uniform near the crack tip. At the same time, because of crack propagation, the structure shows the nonlinear behavior in the dynamic spectral response. What is more, the FWHM changes with the crack propagation. And in this paper, the peak search algorithm is used to search the main wavelength. Then, we choose the -3dB spectral bandwidth as FWHM. A schematic diagram is shown as shown in Figure. 4.

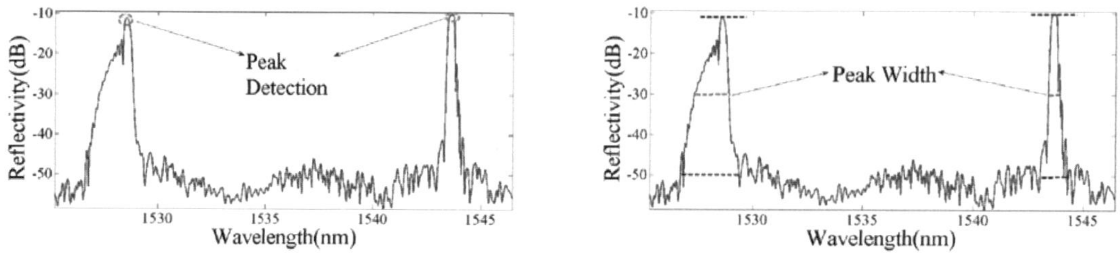

Figure 4. The schematic calculation of FWHM

5. Results

In the monitoring of bridge structure, the research of FBG sensor for uniform axial strain sensing has been fully studied. The measurement principle is based only on the displacement of the Bragg peak reflected by the grating. But, the non-uniform strain form in front of crack singularity will lead to spectrum distortion when the crack growth is detected by FBG sensing.

The paper shows that we get and analyze the simulated spectra which are extracted from FEM and subjected to different strain modes. Simultaneously, based on peak search algorithm and box counting method, crack damage feature extraction method is applied to the analysis of experimental and simulated data. The FWHM damage characteristics of the simulated spectra caused by fatigue crack damage is analyzed. The fatigue crack damage can cause the change of the damage characteristics. And the paper analyzed the FWHM damage characteristics of the experimental and simulated spectra.

The results illustrate that the axial strain distribution and crack propagation of aluminum can be monitored by the FBG sensor. At the beginning of the crack propagation, because of strain changes, the small wavelength shift of the spectrum occurs. In the reflected wave, the spectrum is symmetrical with the uniform wavelength shift. When the crack length increases, the strain gradient of non-uniform along the grating gradually increases.

In contrast, On the contrary, due to the crack propagation into the grating region, FWHM is sensitive to the compression field strain field in front of the crack tip. In this study, FWHM also has an evident correlation with the crack length, in addition to achieving the saturation length of the crack. Figure. 5 shows the relationship between FWHM variation and crack length under simulated damage conditions.

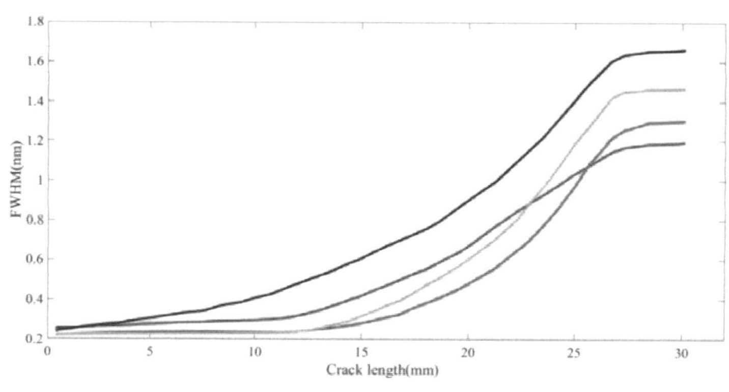

Figure 5. FWHM v.s. crack length for all four parallel-plate

IOP Conf. Series: Materials Science and Engineering **813** (2020) 012002 doi:10.1088/1757-899X/813/1/012002

6. Discussion

There are several forms of strain distribution near the crack singularity, so in the spectrum shape the non-uniform strain field produces the change (i.e. spectrum distortion). Moreover the number of peak splitting increases with the crack through the grating area. At the same time, as in our previous research, the sub peak position detected by grating is in the form of strain.

7. Conclusion

With the strain sensitivity of FBG, the FBG experimental platform is established for monitoring the crack propagation in this paper. And the FBG spectrum changes obviously with the crack propagation, especially the FWHM broadened with the crack length. Meanwhile, the spectra simulation approach is applied for the FWHM extraction, combined with the help of peak seek algorithm. The results show that the FWHM can be used as an effective index for evaluating fatigue crack growth.

8. References

[1] B. Jin, W. Zhang, F. Ren, M. Zhang, W. Dai, and Y. Wang, "Mechanism of subordinate peak skewing of fbg sensor during cracks propagation monitoring on aluminum alloy structure," Journal of Sensors, pp. 1-7, 2017.

[2] K. Bremer, F. Weigand, Y. Zheng, L. S. Alwis, R. Helbig, and B. Roth, "Structural health monitoring using textile reinforcement structures with integrated optical fiber sensors," Sensors, vol. 17(2), pp.345, 2017.

[3] X. Yu, Z. Fan, S. Puliyakote, and M. Castaings, "Remote monitoring of bond line defects between a composite panel and a stiffener using distributed piezoelectric sensors," Smart Material Structures, vol. 27(3), 2018.

[4] R. B. Jenkins, P. Joyce, and D. Mechtel. "Localized Temperature Variations in Laser-Irradiated Composites with Embedded Fiber Bragg Grating Sensors." Sensors vol. 17.2, 2017.

[5] M. A. Xi-Qiang, F. Yang, Y. J Xue, and L. I. Ji-Shun, "On-line monitoring method for fatigue of mechanical structure parts," epee, 2017.

[6] A. Kelly, "An introduction to composite materials," Concise Encyclopedia of Composite Materials, vol. 2(1), pp. xvii–xxix. 2018.

[7] D. Coutellier, "Composites and dynamic failures: experimental and numerical solving strategies and solutions." Solid Mechanics & Its Applications, vol. 192, pp. 611-643, 2017.

[8] A. Rajabzadeh, R. Heusdens, R. C. Hendriks, and R. M. Groves, "Calculation of the mean strain of smooth non-uniform strain fields using conventional fbg sensors." Journal of Lightwave Technology, vol. 1-1, pp.99, 2018.

[9] B. Jin, W. Zhang, M. Zhang, F. Ren, D. Wei, and Y. Wang, "Investigation on characteristic variation of the fbg spectrum with crack propagation in aluminum plate structures." Materials, vol. 10(6), pp. 588, 2017.

[10] S. Cięszczyk, and P. Kisała, "Inverse problem of determining periodic surface profile oscillation defects of steel materials with a fiber bragg grating sensor." Applied Optics, vol.55(6), pp. 1412, 2016.

[11] R. B. Jenkins, P. Joyce, and D. Mechtel, "Localized temperature variations in laser-irradiated composites with embedded fiber bragg grating sensors." Sensors, vol. 17(2), 2017

IOP Conf. Series: Materials Science and Engineering **813** (2020) 012003 doi:10.1088/1757-899X/813/1/012003

Investigation on Temperature Sensitivity of Fiber Bragg Grating Sensor under Global Temperature Variations Experience

Bo Pang[1,a] ,Meng Zhang[2,b*],Yuege Zhou[3,c],Shouwen Liu[4],Weifang Zhang[5]

[1]Beijing Institute of Spacecraft Environment Engineering Beijing Key Laboratory of environmental and reliability testing for aerospace mechanical and electrical products, Beijing, China

[2,5] School of Reliability and Systems Engineering, Beihang University, Beijing, China

[3,4] Beijing Institute of Spacecraft Environment Engineering Beijing Key Laboratory of environmental and reliability testing for aerospace mechanical and electrical products, Beijing, China

[a]1050921989@qq.com,[b]Zhangmeng123@buaa.edu.cn,[c]zhouygb05250@163.com

Abstract. An experiment was developed to relate the wavelength shifts resulting from temperature changes of fiber Bragg grating (FBGs) to the thermal expansion coefficients. The range of the temperature is set from -60 °C to 100 °C as the simulation of outer space temperature environment and the experiment results show that the primary wavelength shift of this sensor due to temperature changing is more precise with the secondary polynomial fitting than the theory relatively linear fitting. Consequently, the reason of the difference between the theoretic and the experiment calculation has been analyzed.

1. Introduction

Fiber Bragg grating (FBG) technology is a promising measurement concept for future sensor system applications. It has been extensively proposed recently and utilized especially in monitoring the aerospace structures and materials health condition [1,2]. Compared with many other sensors, such as PZT [3], ICMS, the FBG [4] sensors have many advantages of being lightweight and flexible. In general, the response of FBGs are determined by coupling effect of strain and temperature, and the primary wavelength shift is linear for a broad range of average strain [5], however, for the temperature response of FBGs the main wavelength changing is much more complicated. There are a lot of researches on FBGs temperature sensitivity based on theory and experiment analysis. Wolfgang Eche et al [6]. proposed a distributed sensor network system consisting 12 FBGs to monitoring the structural health of the X-38 spacecraft undergone the temperature range from -40°C to 200°C and the strain range of -1000με to 3000με.

In this paper, considering the FBGs temperature sensitivity independently in experiment, then the influence of strain effect is excluded, so the FBGs is in the stress-free state. When the FBGs sensors is under cryogenic temperatures, the FBGs temperature sensitivity becomes more nonlinear. Furthermore, the primary wavelength shifting with the temperature increasing per 5°C performs better in secondary polynomial fitting than the theory relatively linear fitting. The paper consists two parts, the section II shows the origination of the experiment design, and the section III shows the detail discussion of the experiment results.

2. Experiment design

In this concept, shown in figure 1, Bragg wavelength-shifts are measured to determine temperature effects that act on the grating section. This study building the FBG temperature sensitivity detection experiment platform, contains two major segments as a temperature control system and an optical sensing system (SM125). To simulate the outer spacecraft temperature circumstances, the rang of temperature is set from -60°C to 110°C and the rate of its growth is 5°C. To acquire the temperature sensitivity, the temperature control of a temperature box, the four FBG sensors are parallel placed in it under unstressed state. Moreover, an optical sensing system (SM125) are used here to get the reflectivity response of the FBG sensor under different temperatures.

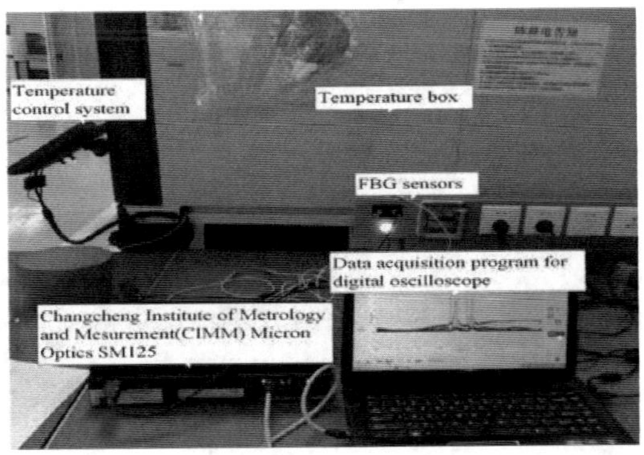

Figure 1. The experiment setup for FBG sensor temperature sensitivity detection system

3. Result and Discussion

The response signal assessed from experiment can be seen in the following figure 2. The lateral axial presents the wavelength and the longitudinal shows the FBG reflectivity spectrum under different temperatures, 0°C,5°C. Obviously, the primary wavelength shift to the longer wavelength with the temperature increasing.

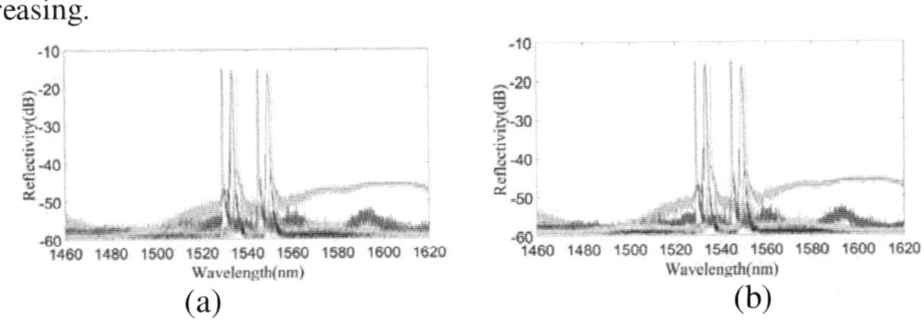

Figure 2. The reflectivity spectrum under temperatures. (a) in the 0°C; (b) in the 5°C.

To analysis the relationship between the temperature characteristic primary wavelength shifting and the temperature changing under extreme environment. The features were obtained by extracting from the FBG reflectivity spectrums under the Temperature range from -60°C to 110°C, and the gradient is 5°C. Then, there are 35 groups experiment data can be processed in this test for each grating position. As mentioned before, the four FBG sensors with eight grating are placed in the temperature box to eliminate the random error. The processed results can be seen in the figure 3, and the axial lateral is the temperature and the longitudinal shows the primary wavelength.

IOP Conf. Series: Materials Science and Engineering **813** (2020) 012003 doi:10.1088/1757-899X/813/1/012003

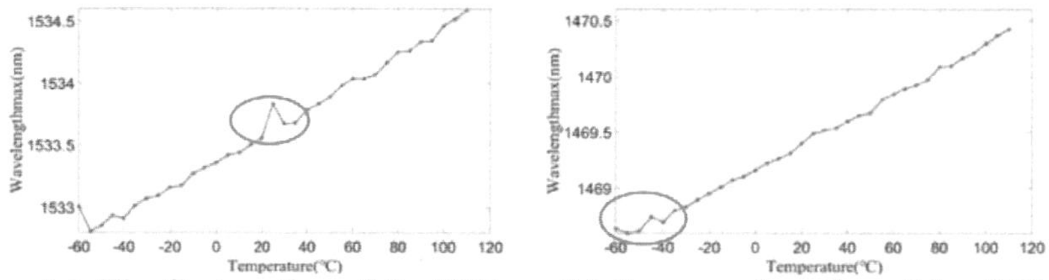

(a) The first grating of the CH1 (b) the second grating of the CH1

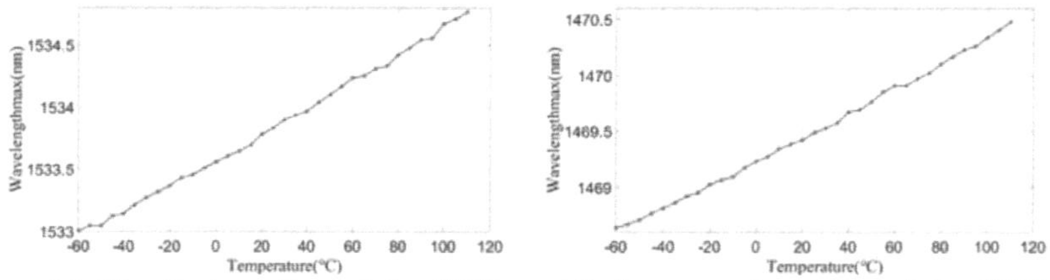

(c) The first grating of the CH2 (d) the second grating of the CH2

Figure 3. The primary wavelength changing under different temperature.

The cycles presented in the figure 3. (a) and (b) clearly shows the singularity in wavelength shifting with temperature increasing. These wavelengths shifting may appear randomly for the reason that the other FBGs do not show the same performance. So, the FBG3 and FBG4 are used to analysis the sensitivity coefficient, what's more, in other parts of the diagram, the variation curve between the primary wavelength shifting and the temperature are not satisfied the one-variable linear regression, shown in figure 4. Then, the coefficient of them may not be equal to the theoretical calculation [7] constant coefficient 0.01nm/°C. The error between the experiment coefficient and the theory is result that when the material of fiber grating expands with heat and contracts with cold [8], it appears unusual that the fiber optical and mechanical properties changed experienced different temperature stages.

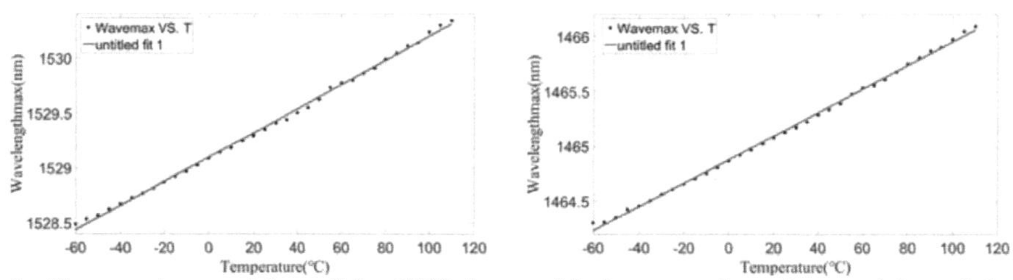

(a) the first grating position of the CH3 data (b) the second grating position of the CH3 data

Figure 4. Linear regression fitting results between primary wavelength and the temperature in CH3

Table 1. The function and the evaluation indicators of the fitting results

Grating position 1	Grating position 2
F (x) = 0.01074x + 1465	F (x) =0.01075x + 1466
SSE： 0.0176	SSE： 0.0177
R-square： 0.9982	R-square: 0.9983
RMSE： 0.02315	RMSE: 0.02316

In the following, we operate the linear and quadratic fitting between the primary wavelength and the temperature. To evaluate the performance of two fitting methods, the three indicators [9] such as, the sum of squares due to error (SSE), Coefficient of determination(R-square) and the Root mean squared error (RMSE) are chosen here to quantitative the fitting performance. The fitting results diagram show the quadratic model fits more successfully than the linear model, shown in figure 5, and the details can be seen in the following tables 1 (CH3) and table 2 (CH4).

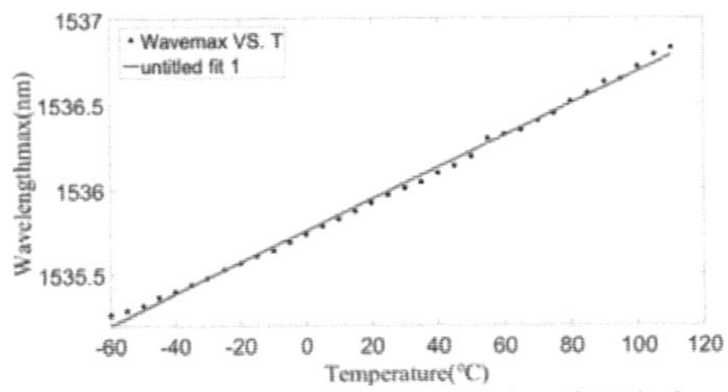

(a) the first grating position of the CH4 data

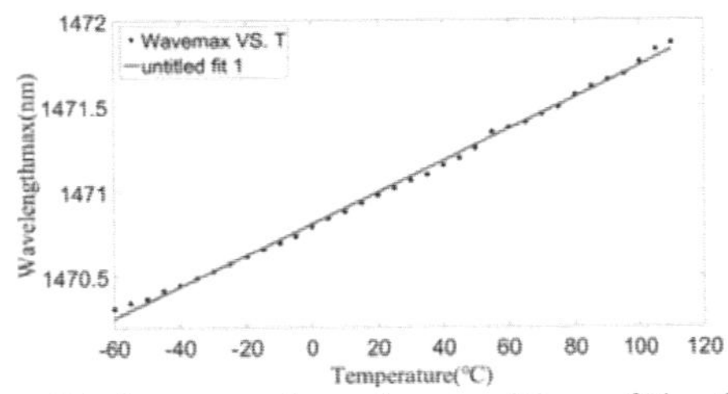

(b) the second grating position of the CH4 data

Figure 5. Linear regression fitting results between primary wavelength and the temperature in CH4

Table 2. The function and the evaluation indicators of the fitting results

Grating position 1	Grating position 2
F (x) = 0.00923x + 1472	F (x) = 0.00922x +1471
SSE: 0.01931	SSE: 0.01932
R-SQUARE: 0.9974	R-square: 0.9975
RMSE: 0.0241	RMSE: 0.0242

The tables show the linear regression result of the measured Bragg wavelength versus temperature. The theory temperature sensitivity of the FBG is 0.01nm/°C, however the experiment data fitting results are not equal to it with shifting left and right side. Moreover, by calculating the relative error, the results as follows:

$$\delta_2 = \frac{0.01075-0.01}{0.01} \times 100\% = 7.5\% \tag{1}$$

$$\delta_3 = \frac{0.01-0.00923}{0.01} \times 100\% = 7.7\% \tag{2}$$

Moreover, the temperature ranging from 45°C to 100°C, the coefficients is not following the linear calculating result. When the temperatures are below -40°C, the curve is almost nonsmoothed, i.e., showing fluctuate temperature sensitivity.

In order to choose the beat fitting method and to solve the singularities in the prementioned linear fitting. The quadratic fitting method is proposed here to compare with the theory ideal method. The processing result about CH3 (shown in figure 6) and CH4 (shown in figure 7)can be seen in the following table 3 (CH3) and table 4 (CH4).

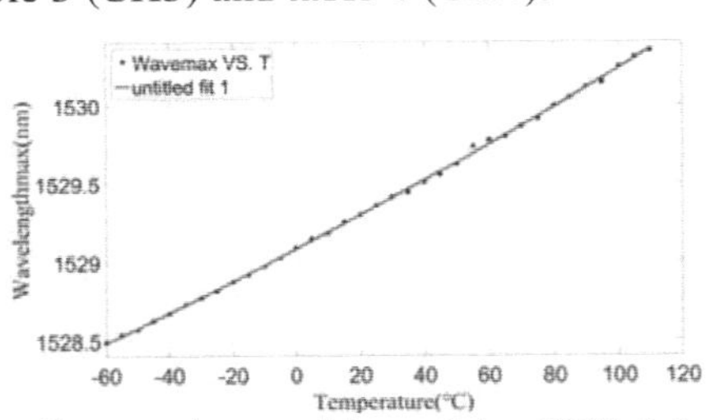

(a) the first grating position of the CH3 data

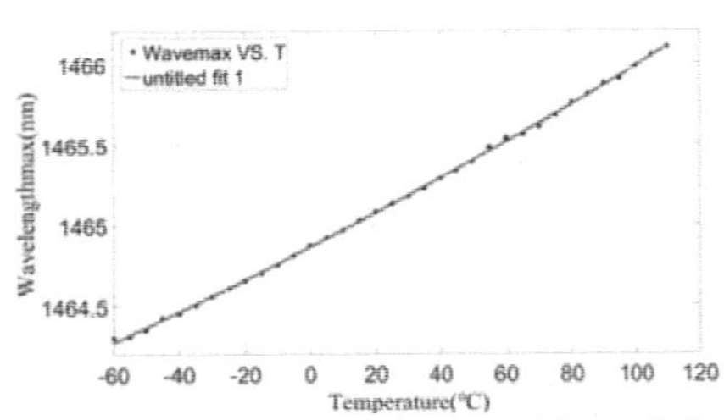

(b) the second grating position of the CH3 data

Figure 6. The quadratic regression fitting results between primary wavelength and the temperature in

CH3

Table 3. The function and the evaluation indicators of the fitting results

Grating position 1	Grating position 2
$\mathbf{F(X) = 0.000009514 * X^2 + 0.01055 * X + 1529}$	$F(X) = 0.000009513 * X^2 + 0.01056 * X + 1529$
SSE: 0.006015	SSE: 0.006016
R-square: 0.9994	R-square: 0.9993
RMSE: 0.01371	RMSE: 0.01372

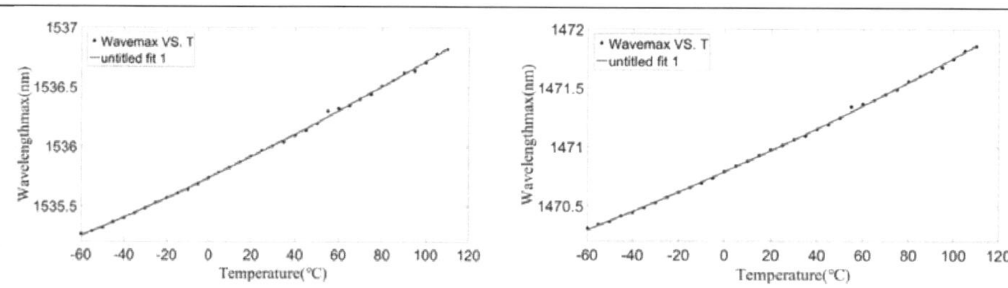

(a) the first grating position of the CH4 data (b) the second grating position of the CH4 data

Figure 7. The quadratic regression fitting results between primary wavelength and the temperature in CH4

Table 4. The function and the evaluation indicators of the fitting results

Grating position 1	Grating position 2
F (x) = 7.132e-06$* \boldsymbol{X}^2$ +0.01016 $* \boldsymbol{X}$ + 1534	F (x) = 7.131e-06$* X^2$ +0.01017 $* X$ + 1534
SSE: 0.009737	SSE: 0.009735
R-square: 0.999	R-square: 0.999
RMSE: 0.01744	RMSE: 0.01743

In the above discussion about the theoretical linear fitting method and the quadratic fitting methods for CH3 and CH4 data processing. Comparing the three indexes proposed here, the quadratic method has a better performance than the theoretical method. Meanwhile, according to the linear fitting of the wavelength offset and the temperature in curve, it is found that the linear relationship between them is an approximate function, and the constant temperature sensitivity coefficient represents the average of the temperature sensitivity coefficient in a certain temperature range. When the temperature increases over 45°C, the temperature sensitivity coefficient of FBG increases quickly along with the gradually rise of temperature, then, the quadratic relation used to fit the relative primary wavelength drift and temperature has a better effect than linear fitting method.

The reason for the difference between the theoretical calculation value and the experiment value in FBG temperature sensitivity is that the theoretical values are obtained by the mechanical parameters of the fiber grating material. However, in experiment, the original mechanical parameters such as the refractive index and the temperature coefficient have changed after under the temperature difference alternately. Then the temperature sensitivity coefficient which mostly based on the above two fiber parameters will be changed. In practice, if the FBG sensors are used to accurately measure temperature, it is not appropriate to calculate the FBG temperature sensitivity coefficient by using the

FBG material mechanical parameters, moreover, the specific temperature sensitivity coefficient should be determined based on experiment.

4. Conclusions

In this paper, the FBG response signals under different temperature values have been processed to define the temperature sensitivity of the sensor. Comparing the theoretical linear fitting methods, the quadratic fitting model performs better that have the smallest SSE and RMSE, what's more, the R-square is the biggest. Through analysis the error between the linear fitting and the experiment quadratic fitting results, the reason for the difference is that the FBG materials expands with heat and contracts with cold when the fiber grating sensing the temperature gradient. And the FBG reflectivity signal has been changed with the FBG parameters refractive index and the temperature coefficient changed.

Acknowledgements

The authors wish to thank the editor and anonymous referees for their constructive comments and recommendations, which have significantly improved the presentation of this paper. This study is supported by National Defense Science and Technology Bureau Quality and Reliability Project(No. JSZL2017203B007) and Beijing Natural Science Foundation - Rail Transit Joint Fund Project(No. L161010) and Technical basic research project(No. JSZL2017601C002).

5. References

[1] Chai Q, Tian Y, Zhang J Z, et al. Spectral characteristics of rotated fused polarization maintaining fiber Bragg gratings subjected to transverse loading[C]// International Conference on Optical Fiber Sensors. 2017:103233Q.
[2] Cruse T A, Besuner P M. Residual life prediction for surface cracks in complex structural details[J]. Journal of Aircraft, 2012, 12(12):369-375.
[3] Ogisu T, Okabe Y, Takeda N. Development of damage monitoring system for aircraft structure using a PZT actuator/FBG sensor hybrid system[J]. 2004, 5388(10):53-54.
[4] Okabe Y , Tanaka N , Mizutani T , et al. Effect of fiber coatings on detection of transverse cracks in CFRP composites using FBG sensors [J]. Icf10 Honolulu, 2013.
[5] Chen Y, Liu B, Liu H, et al. Load location and measurement system based on fiber Bragg grating sensor[J]. Journal of Mechanical Engineering, 2016.
[6] Ecke W, Grimm S, Latka I, et al. Optical fiber grating sensor network based on highly reliable fibers and components for spacecraft health monitoring[J]. Proc Spie, 2001, 4328:297-305.
[7] Wu M C. Pressure effects on the temperature sensitivity of fiber Bragg gratings[C]// American Institute of Physics, 2013:1570-1577.
[8] Abtahi M, Simard A D, Doucet S, et al. Characterization of a Linearly Chirped FBG Under Local Temperature Variations for Spectral Shaping Applications[J]. Journal of Lightwave Technology, 2011, 29(5):750-755.
[9] Tong H. Convergence rates for empirical bayes estimators of parameters in linear regressin models[J]. Communications in Statistics, 1996, 25(6):1325-1334.

IOP Conf. Series: Materials Science and Engineering **813** (2020) 012004 doi:10.1088/1757-899X/813/1/012004

Numerical Analysis on Target Strength of the Filled Spherical Elastic Shell

Bing Jia[a], Jihui Wang[b], Guijuan Li[c], Yunfei Chen[d]

Science and Technology on Underwater Test and Control Laboratory, Dalian, Liaoning, China

[a]jbrobin@sina.com, [b]wangjihuihgc@163.com, [c]18641198261@163.com, [d]yunfeidlut@163.com

Abstract: In this paper, the acoustic target strength of spherical shells in water with internal fillers is calculated by using the normal series expansion method, and the effects of metal and non-spherical shell materials, size factors and different fillers on the target strength of spherical shells are analyzed. The results show that for different elastic shell parameters, the influence of the acoustic characteristics of the internal filler on the target strength is quite different. For non-metallic sound transmission spherical shells filled with fluid, only when the sound velocity difference between the filler and the water medium is large, the target strength is greatly improved compared with the corresponding rigid sphere, and the shell size parameters have little effect on the target strength. However, for metal spherical shells with internal fillers, the material and size factors of the shell have a greater impact on its target strength, and the sensitivity to sound velocity mismatch inside and outside the shell is relatively small.

1. Introduction

The prediction of target intensity can be understood as solving the scattered acoustic field in three-dimensional fluid space, which satisfies the surface boundary conditions, wave equation and radiation conditions. In theory, target intensity prediction is to solve a mathematical and physical problem, which is essentially a problem of acoustic scattering. At present, the theoretical solutions of acoustic scattering problems mainly include integral equation method, Rayleigh normal series solution caused by separation of variables method, Sommerfeld-Watson integral transformation method[1], resonance scattering theory and singular point expansion theory.

Rayleigh was the first scholar to study the scattering of a sphere. He proposed an approximate solution to the scattering sound field when the wavelength is much larger than the diameter of a sphere[2]. Junger first used the thin shell theory to study the acoustic scattering of air-filled elastic shells[3]. Goodman and Stern have studied the plane wave acoustic scattering of submerged spherical shells with precise elastic mechanics theory, and derived Rayleigh normal series solution of scattering sound field of spherical shells filled with fluid in the same fluid[4]. J.A. Fawcett gave the solution of acoustic field of elastic shell partially filled with fluid[5]. Gregory Kaduchak and Charles M. Loeffler studied the relationship between material parameters and target strength of liquid filled spherical shells in water, and gave some test results [6]. In this paper, the separation variable method is used to calculate the target strength of spherical shells in internal and external fluid media, and the effects of the acoustic characteristics of metal spherical shells, non-metallic sound transmission spherical shells, dimensions and different fillers on the target strength of spherical shells are calculated and analyzed. The research of this paper will serve the design of underwater acoustic scattering body.

2. Computing method

The method of separating variables is a classical method for solving wave equation. The scattered wave is expressed as Rayleigh normal series solution of basic wave function. The expansion coefficients of series are obtained according to boundary conditions[7] and incident wave, and then the scattered wave solution is obtained.

2.1. Normal mode solutions of target strength of rigid sphere

The exact solution of rigid spherical scattering is often regarded as a standard problem. Scattering field of rigid sphere obtained from expansion coefficient:

$$\phi_s(r,\theta) = -\sum_{n=0}^{\infty} i^n (2n+1) \frac{j_n'(ka)}{h_n^{(1)'}(ka)} P(\cos\theta) h_n^{(1)}(kr) \tag{1}$$

Definition of shape function:

$$f(x,\theta) = \frac{2r}{a} e^{-ikr} \frac{\phi_s(x,\theta)}{\phi_i(x)} \tag{2}$$

According to the definition of target strength in far field, the relationship between the shape function of the sphere and the target strength can be obtained, as shown below:

$$\begin{aligned} TS &= 20\log\left|\frac{a}{r} f(x,\theta)\right| \\ &= 20\log\left|\frac{a}{r\cdot x}\sum_{n=0}^{\infty}(2n+1)\frac{j_n'(x)}{h_n^{(1)}(x)} P_n(\cos\theta)\right| \end{aligned} \tag{3}$$

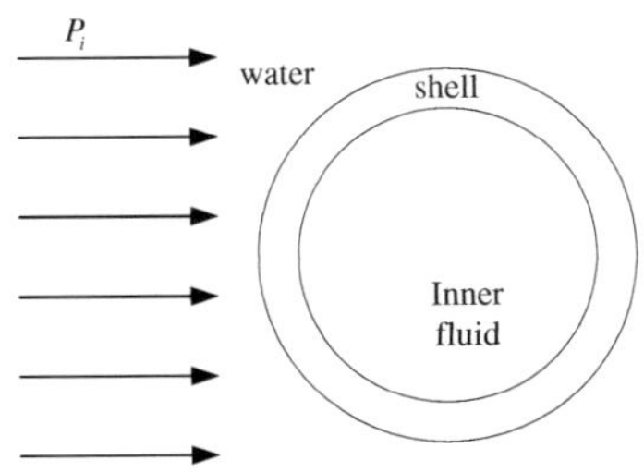

Figure 1. The acoustic scattering of elastic sphere shell

2.2. Normal mode solutions of target strength of elastic sphere shell

There is an elastic spherical shell immersed in fluid, as shown in Figure 1. The outer radius of the spherical shell is r, the thickness is h. The velocity of compression wave is represented as Cl, and the velocity of shear wave is represented as Ct.

When plane waves incident on a spherical shell, the expression of incident plane waves in polar coordinates is that the wave equation satisfies the boundary conditions at the spherical boundary, solve the equation， the expression of scattered sound field is as below[8]:

$$P_s(r,\theta) = P_0 \sum_{n=0}^{\infty} (2n+1) j^n B_n h_n^{(2)}(kr) P_n(\cos\theta) \tag{4}$$

The far-field sound pressure expression is as follows:

$$P_s(r,\theta) = P_0 [\sum_{n=0}^{\infty} (2n+1) j^{2n+1} B_n P_n(\cos\theta)] \frac{e^{-jkr}}{r} \tag{5}$$

IOP Conf. Series: Materials Science and Engineering **813** (2020) 012004 doi:10.1088/1757-899X/813/1/012004

When the far-field sound pressure is reduced to a distance of 1 m from the center of the spherical shell, the target strength of the spherical shell is expressed as follows:

$$TS = 20\log\left|\frac{P_s(r,\theta)}{P_i(r,\theta)}\right|_{r=1m} \tag{6}$$

3. Simulation calculation and analysis

3.1. Calculation mode

In the simulation, we only considered the transmission and scattering of sound waves in spherical shells and filling materials. The interaction between the two media has less effect on sound waves and is not considered in the calculation. The outer radius of the spherical shell is 0.5m. The inner part is filled with liquid and the outer part is water medium. As shown in Table 1, the acoustic characteristics parameters of metal material aluminum and non-metal material natural rubber are respectively shown, while those of air, water, brine, alcohol and carbon tetrachloride are shown in Table 2. In the simulation, the parameters used are the frequency range of 50 Hz-10 kHz.

Table 1. Acoustic parameters of spherical shell materials

Materials	Cl[m/s]	Ct[m/s]	Density [kg/m^3]
Aluminum	*6260*	*3080*	*2700*
gum rubber	*1418*	*210*	*970*

Table 2. Acoustic parameters of fillers in spherical shells

Fluid	Cl[m/s]	Density [kg/m^3]
Air	*331*	*0.0013*
Water	*1500*	*1000*
Saline water	*1702*	*1131*
Alcohol	*1168*	*789*
CCL4	*935*	*1590*

3.2 Calculation results and analysis

When the coordinate origin is in the center of the circle and plane wave is incident, the field point is located 1000 meters away from the center of the sphere(monostatic).

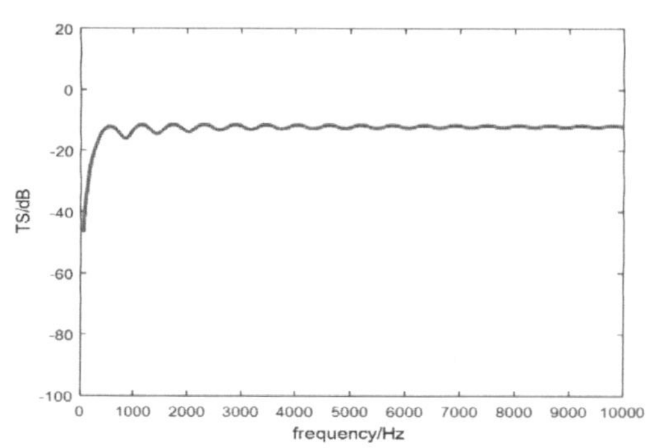

Figure 2. Target strength of a rigid sphere of radius 0.5 m

As shown in Figure 2, when the radius of the rigid sphere a is 0.5m and Ka >>1, K is the number of sound waves, the target strength is about -12dB. Using the method above, the spherical shell material fixed to aluminum, we change the filling material and calculate TS, the results are shown in Figure 3.

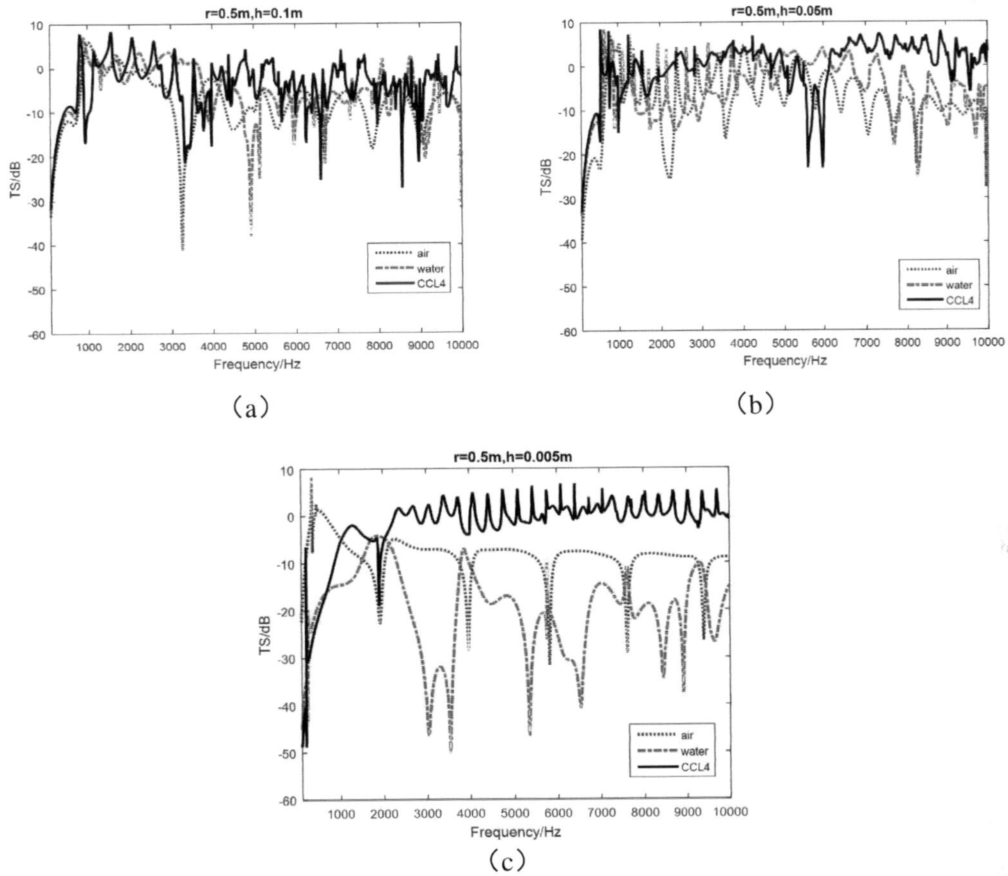

Figure 3. Target strength of three different internal fillers for Aluminum Spherical Shells

As shown in Figure. 3 (a), there is no significant difference in the effect of target strength enhancement when air, water and CCl4 medium are filled in the aluminum spherical shell with a radius of 0.5m and a thickness of 0.1m.

As shown in Figure. 3 (b), when the aluminum spherical shell with a radius of 0.5m and a thickness of 0.05m is filled with ccl4, the target intensity near 6kHz is higher in the whole frequency band because of the great difference between the sound speed and water. Compared with the rigid sphere with the same size, the maximum enhancement is about 20dB in the broader band. When water and air are filled in the interior, the target strength fluctuates sharply, the elastic echo of aluminum spherical shell is obvious, and the target strength is greatly enhanced at a specific frequency

As shown in Figure. 3 (c), when the aluminum spherical shell with a radius of 0.5m and a thickness of 0.005M is filled with ccl4, the target strength is relatively high in the 3-10kHz and is about 0dB in the broader band, which is obviously enhanced compared with the rigid sphere with the same size. At frequencies below 1 kHz, when the internal filler is air, the target strength increases greatly, but only slightly at other frequencies. For the internal filling water medium, it has no effect on target strength enhancement.

By comparing Figures 3 (a, b, c), it can be found that the difference of sound velocity between internal and external fluid is not significant for thick metal spherical shells. When the shell becomes thinner, the mismatch of sound velocity between internal and external fluid has a greater effect on target strength enhancement.

IOP Conf. Series: Materials Science and Engineering **813** (2020) 012004 doi:10.1088/1757-899X/813/1/012004

The spherical shell material fixed to rubber, we change the filling material and calculate the TS, the results are shown in Figure 4.

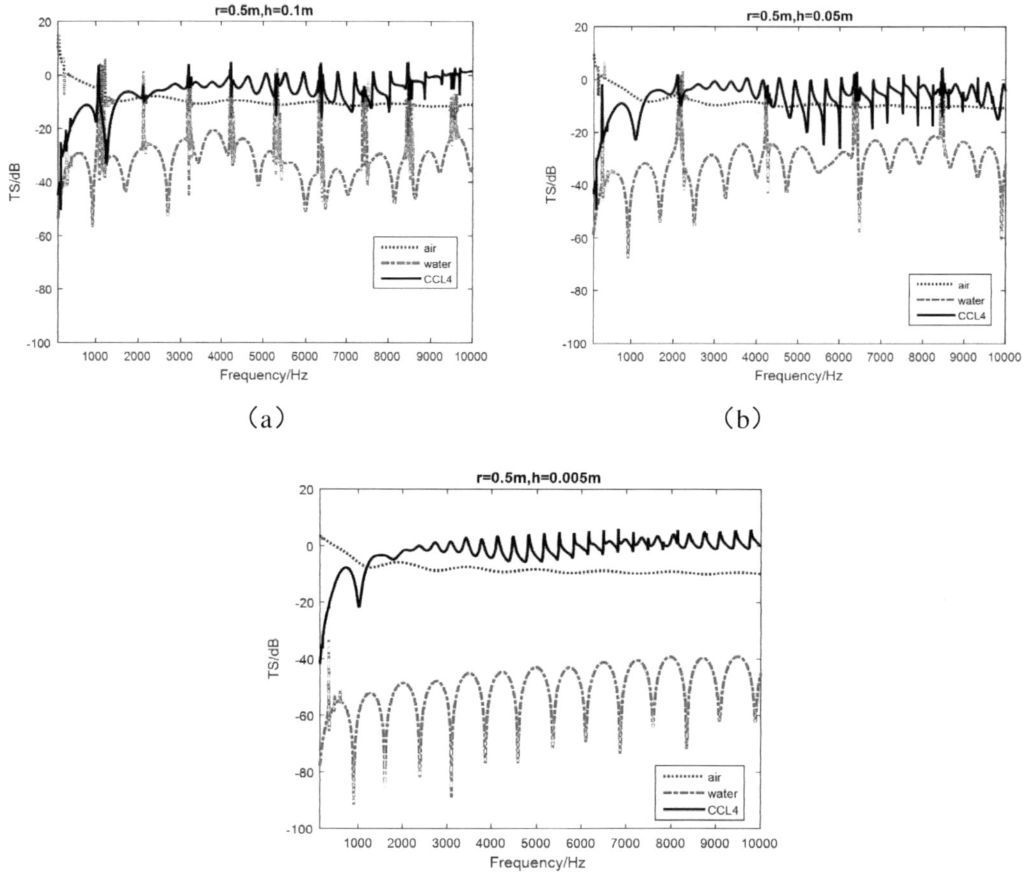

(a) (b) (c)

Figure 4. Calculations of Target Strength of Natural Rubber Spherical Shells with Three Different Internal Fillers

As shown in Figure. 4 (a), the target strength of natural rubber spherical shell with radius r=0.5 mm thick h=0.1 m filled with CCl4 and air medium is higher than that filled with water medium except for 9 frequencies, which indicates that the sound velocity mismatch effect of internal and external fluid medium is obvious. When air is filled in the interior, the effect of internal gas is remarkable. The target strength fluctuates very little in a wider band, and the target strength is enhanced compared with the rigid spherical shell.

As shown in Figure. 4 (b), the target strength law of natural rubber spherical shell with radius r=0.5m thick h=0.05m is basically consistent with that of Figure. 4 (a). As shown in Figure g. 4 (c), the target strength of natural rubber spherical shell with radius of 0.5m and thickness of 0.005M is very small when filled with water, which is due to the good sound transmission performance of natural rubber. Compared with Figure 3, the target strength curves of natural rubber spherical shells filled with three kinds of fillers are basically the same, which shows that the matching degree of sound velocity of fluid inside and outside the shell plays a major role in the sound transmission of non-metallic sound transmitting material.

To compare the effects of the shell material on TS, we calculated TS of spherical shells with the same shape structure and filled liquid, but different materials, as shown in Figure 5 and Figure 6.

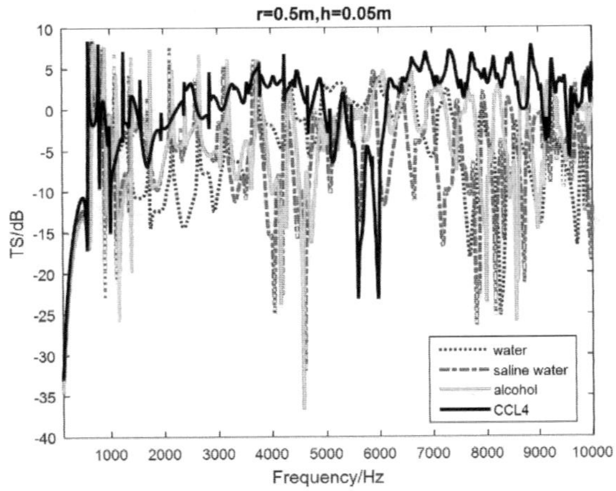

Figure 5. Target Strength of Aluminum Spherical Shells Filled with Four Kinds of Liquids

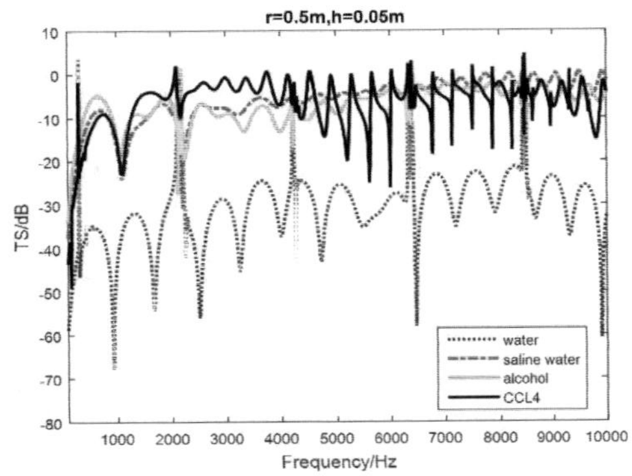

Figure 6. Target Strength of Natural Rubber Spherical Shells Filled with Four Kinds of Liquids

From Figure. 5 and 6, it can be seen that the fillers in the spherical shell are different fluids, the density of these fluids is different, and the sound velocity in the fluid is different, which has an effect on the sound scattering. The mismatch of the sound velocity inside and outside the spherical shell plays an important role in improving the target strength of the spherical shell. Due to the size factor of shell material, the variation law of target strength is different, which should be fully considered in practical engineering application.

4. Conclusions

In this paper, the effects of the material, size and acoustic characteristics of different fillers on the strength of spherical shells are calculated and analyzed. The following conclusions are drawn:

(1) The difference of sound velocity between inner and outer fluid of metal thick spherical shell is not significant, but the mismatch of sound velocity between inner and outer fluid of metal thin spherical shell plays a greater role in enhancing its target strength.

(2) The matching degree of sound velocity of fluid inside and outside the shell plays a major role in the target strength of non-metallic sound transmission materials.

In addition, parameters such as the microscopic structure of the spherical shell and the temperature of the filling material also affect the TS value. These contents will be studied later.

5. References

[1] Kevin L. Williams and Philip L. Marston. Backscattering from an elastic sphere: Sommerfeld-Watson transformation and experimental confirmation. J.Acoust.Soc.Am.79(3), 1093-1102, 1985

[2] L.Rayleigh. Theory of Sound. Dover Publieations, NewYork, 1945
[3] M.C.Junger. Sound scattering by thin elastic shells.J.Acoust.Soc.Am. 24(4):336-373, 1952
[4] R.R.Goodman and R.Stern. Reflection and transmission of sound by elastic spherical shells J J.Acoust.Soc.Am. 34(3):338-344, 1962
[5] J.A.Fawcett. Scattering from a partially fluid-filled,elastic shelled sphere.J.Acoust.Soc.Am. 109(2):508-513, 2001
[6] G. Kaduchak and C. M. Loeffler. Relationship between material parameters and target strength of fluid-filled spherical shells in water:Calculations and observations. *IEEE J. Ocean. Eng.*, vol. 23, no. 1,pp. 26-30, Jan, 1998.
[7] Gaunaurd G.C.,Werby M.F.Sound scattering by resonantly exited, fluid-loaded, elastic spherical.J.Acoust.Soc.Am.,90(5):2536-2550,1991
[8] David M. Deveau and Anthony P. Lyons. Fluid-Filled Passive Sonar Calibration Spheres:Design, Modeling, and Measurement. IEEE Journal of Oceanic Engineering, VOL. 34, NO. 1, 2009

IOP Conf. Series: Materials Science and Engineering **813** (2020) 012005 doi:10.1088/1757-899X/813/1/012005

Computer Modeling on the Effect of the Multistep Nanoimprinting Process in Elastic Deformation Recovery of COC (Cyclic Olefin Copolymer)

Hamzah Almadani [1,2], X Chen[1,3*] and Junfei Cheng[1]

[1] Faculty of Mechanical Engineering and Mechanics, Ningbo University, Ningbo 315211, Zhejiang Province, P.R.China
[2] Mechanical Engineering Dept. Faculty of Engineering- Sana'a University. P.O.Box: 12544- Sana'a - Yemen
[3] Zhejiang Provincial Key Lab of Part Rolling Technology, Ningbo University, Ningbo 315211, P. R. China

chenxing@nbu.edu.cn

Abstract. The effect of the multistep imprint in elastic deformation recovery was studied numerically with more than one roll and different depth ratios. As well as effect hold-on time on the elastic deformation recovery of polymer Cyclic Olefin Copolymer (COC) material was studied using Generalized Maxwell (GM) constitutive model and their parameters in finite element simulation software of ABAQUS were presented. Firstly, the modified Genetic Algorithm (GA) was employed to obtain a Prony series of the viscoelastic model in terms of experimental data of the complex modulus for COC. To evaluate the elastic recovery of the microstructure, the cross-sectional area of micro-channel before and after hot-imprinting was computed and compared numerically. The simulated results were analysed and clearly showed an improvement in the elastic recovery rate of the microstructure area as the multi-step of rolling was employed.

1. Introduction

Since 1996, several attempts have been subsequently made to study the methods of polymer flow during the imprinting process as **Chou** developed nanoimprint lithography [1]. Hot roll-to-plate imbossing process is a continuous forming method that is a promising means of forming the microstructures required in medical devices or so on. In the hot roller press forming process, the polymer material is firstly preheated to the forming temperature. The polymeric plate is pressed initially by rotation movement of rolling mold. Then, the cooling of the polymeric substrate is conducted before demolding. Finally, the load is then removed and the part separated from the mold (demolding).

Numerical methods were employed to simulate polymer flow during imprinting by **Hirai** and **Song** [2, 3]. **Young** analysed polymer flow using a viscous fluid model to predict the polymer flow behaviour during the imprinting [4]. Nanoimprint flow behaviour has been modelled by the coarse-grain method derived from the 3D Navier-Stokes equations, Galerkin finite element program and using MARC software based on the Moony-Rivlin model by **Sirotkin**, **Rowland** and **Scheer** sequentially [5,6,7]. The deformation behaviour of solid polymer was observed by two methods, synchronous observation and asynchronous analysis by **Liu** [8]. The mathematical non-linear model of the hot imprint process included three steps of heating, imprinting and demolding was done by

Narijauskaitė [9]. Rheological analysis of COC with application to hot imbossing for microfabrication made by **Jena** [10]. In 2018 **Hongbo** provides a reference and direction for the further explorations and studies of large-area micro/nanopatterning technologies [11].

The most famous process used in the hot imbossing process is the isothermal hot imbossing which the mold and polymer will be heated to the same temperature [8]. When the mold is released, it gradually separates from the surface of the sheet, the binding force on the polymer surface is reduced and the polymer material also experiences a phenomenon of volume expansion that will also lead to elastic deformation of the molded microstructure. Springback phenomenon in the NIL process is studied with the MD simulation and the mold geometry effect on Springback by **Yang** [12]. In 2015 **Qu** studied the effect of mesh on Springback in 3D finite element analysis of Flexible Microrolling [13].

The study of the constitutive model of viscoelastic materials has been a long time, but the few study of dynamic viscoelasticity to determine the discrete relaxation time spectrum by the Generalized Maxwell model and the effect of nanoimprinting parameters process in elastic deformation recovery are reported. The key to accurately describing the characteristics of viscoelastic materials in the process of forming simulation is the accuracy and reliability of the material parameters. In this paper, firstly the parameters of the viscoelastic material model were fitted at the different temperatures of 100°C、 110°C、 120°C、 13°C、 and 140°C. Then the master curve of the TOPAS®COC material model was obtained in terms of the Williams-Landel-Ferry (WLF) shift function at a specific temperature. Then, the ABAQUS finite element analysis software was used to simulate the rolling hot-press molding. Finally, effect of multistep imprint and hold-on time on elastic deformation recovery was studied.

2. A viscoelastic model for the TOPAS® COC resist and Optimization of model parameters

The mechanical properties of polymer materials are both viscous and elastic at the same time in any case. When the temperature of the polymer body is above the glass transition temperature, its state status lays between the ideal elastomer and the ideal viscous body. Under these circumstances, the total stress includes the elastic and viscous stress as it is subjected to an external force, which can be described in viscoelastic model. In our study, the complex modulus based on the viscoelastic generalized Maxwell constitutive model is used to solve the model parameters of the material. For a viscoelastic polymer, the stress relaxation behaviour can be represented by the generalized Maxwell model with N Maxwell units (a spring and a dashpot in series) in parallel with an isolated spring, as shown in figure 1. And the Prony equation of the storage modulus $E_s(\omega)$ and the loss modulus $E_l(\omega)$ of the generalized Maxwell model in the frequency can be expressed as equation (1)[10].

$$E_s(\omega) = E_\infty + \sum_{i=1}^{n} E_i \frac{\omega^2\tau_i^2}{1+\omega^2\tau_i^2} \quad and \quad E_l(\omega) = \sum_{i=1}^{n} E_i \frac{\omega\tau_i}{1+\omega^2\tau_i^2} \tag{1}$$

where ω is the angular frequency domain. E_i and τ_i are relaxation modulus and time constant of the i-th element in the generalized Maxwell model, respectively. If g_i is a dimensionless modulus which can be expressed as: $g_i=E_i/E_0$ and E_0 is the instantaneous modulus which can be expressed as:

$$E_0 = \frac{E_\infty}{(1-\sum_{i=1}^{n} g_i)} \tag{2}$$

where E_∞ is the equilibrium value of E (t) after the time t goes to infinity, now $E_s(\omega)$ and $E_l(\omega)$ also can be expressed as following equation (3).

$$E_s(\omega) = E_0[1-\sum_{i=1}^{n} g_i] + E_0 \sum_{i=1}^{n} \frac{g_i\omega^2\tau_i^2}{1+\omega^2\tau_i^2} \quad and \quad E_l(\omega) = E_0 \sum_{i=1}^{n} \frac{g_i\omega\tau_i}{1+\omega^2\tau_i^2} \tag{3}$$

According to the storage modulus $E_s(\omega)$ and the loss modulus $E_l(\omega)$, the complex modulus E* can be obtained as :

$$|E^*| = \sqrt{E_l^2 + E_s^2} \tag{4}$$

Based on the reduced time concept, the relationship between modulus at temperature T and T_0 can be expressed as equation (5).

$$E(T_0, t) = E(T, t_0 . a_T) \tag{5}$$

Using WLF equation, the time reduction factor for materials can be expressed as equation (6).

$$\log a_T = \frac{-C_1(T-T_0)}{C_2+(T-T_0)} \quad (6)$$

where C_1 and C_2 material constants at reference temperature T_0 [14]. The glass transition temperature Tg=80°C of TOPAS®RCOC was selected as the reference temperature, and the curves at 100°C、110°C、120°C、130°C、and 140°C were translated to form master curve at 80°C, as shown in figure 2.

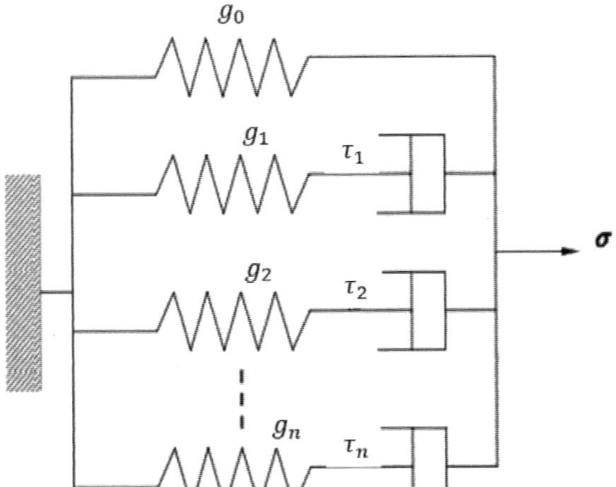

Figure 1. Generalized Maxwell model

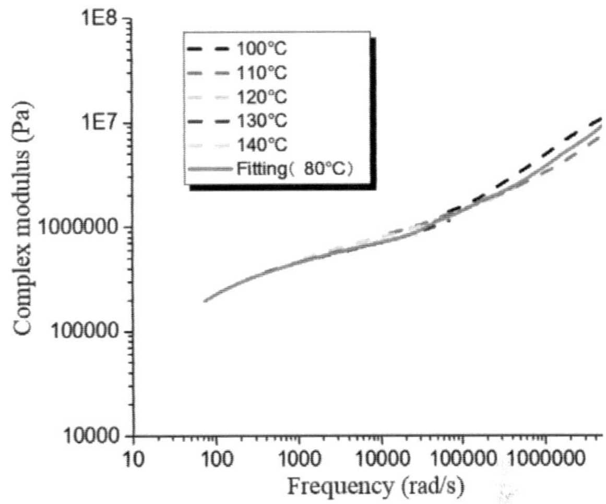

Figure 2. Complex modulus curve at 80°C

The viscoelastic model parameters were obtained by a genetic algorithm at the reference temperature (80°C). In table 1 fitting results are compared with the results at a reference temperature, and comparing results show that the fitting results are in good agreement with the experimental results.

Table 1. Dynamic viscoelastic model parameters obtained by genetic algorithm at 80°C [15]

Prony series	Parameters value	
n	τ_i	E_i
1	1.00E-07	1.53E+07
2	1.00E-06	8.97E+05
3	1.00E-05	7.86E+05
4	1.00E-04	1.68E+05
5	1.00E-03	2.11E+05
6	1.00E-02	9.61E+04
7	1.00E-01	8.07E+04
8	1.00E+00	3.23E+04
9	1.00E+01	5.27E+04
10	1.00E+02	1.62E+04

Note： E_0=1.8E+1MPa

3. Simulations of the model in roll forming of Polymer TOPAS®COC

In the simulation, the mold is often defined as an analytical rigid body, and the polymer substrate is defined as a deformable body. This is because the mold is usually made of metal and its strength is far greater than the strength of the polymer material. In the paper, the polymer sheet is a rectangle with a 100 mm in length dimension and 1.5 mm in the thickness dimension.

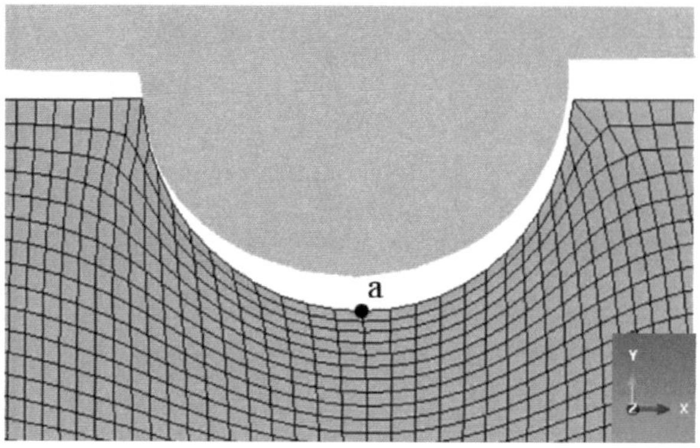

Figure 3. Effect of die on a film substrate

In this paper, the polymer substrate was considered to be viscoelastic material and the basic physical parameters of polymer in tables 1-3. Roll forming is a combination of external force and boundary constraints. A constraint in the X, Y, and R3 directions is set on the bottom edge of the polymer sheet. In the analysis model, a displacement in the Y direction is first set at the reference point of the rolling mold centre. Then, the rolling mold starts to rotate and the thin plate is purely rolling. Constant rotational speed in the Z direction and a constant linear velocity in the X direction are set at the central reference point. The temperature field of the rolling die and the polymer sheet is defined in a predefined field. The transfer process is in a constant temperature environment, ignoring the heat transfer process. In this analytical model, the polymer sheet mesh type is CPE4RT and the total number of grids is 23872 see figure 3.

Table 2. Material physical parameters of TOPAS®COC

Parameters	Values
Density（10^3kg/ m^3）	1.01
Elastic modulus（MPa）	2600
Poisson’s ratio	0.3

Using the principle of translation, the complex modulus data at different temperatures are transformed by the shift factor tabulated in table 3.

Table 3. WLF equation material constants

Parameters	Values
C_1	7.2
C_2	88.5
T_0	150 C°

4. Results and Discussion

In this paper, the elastic recovery rate indicates the change of the elasticity of the microstructure, and the elastic recovery rate (φ) can be expressed as equation (7).

$$\varphi = \frac{A - A_R}{A} \quad (7)$$

where A and A_R are a microstructure cross-sectional area before and after the rebound, respectively.

4.1 Effect of the number of printing steps on recovery deformation of micro-structure

To study the effect of multi-step press on recovery deformation of micro-structure, there are several factors taken into account the imprint depth at each step and the cooling time. When the printing is done using two or three stages through three rollers. The polymer is firstly compressed by a certain percent of imprint depth by the first roller and then roller releases the flat polymeric substrate which it continues to move to a specific position pressed by the second roller to the required depth for two-step process, as shown in figure 4.

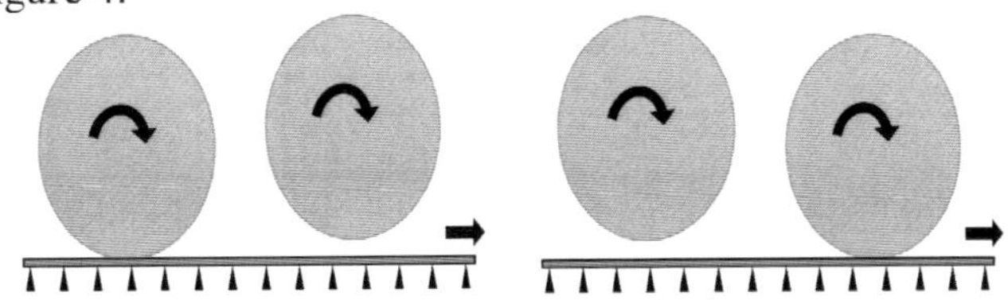

Figure 4. Multistep nanoimprinting process

The simulated imprint results of elastic recovery rate (φ) for three different processes of one-step, two-step and three-step tabulated in table 4. By comparison of these results of 5.60% in one step imprint process, 2.98% in two-step process and 2.00% in the three-step process, it is demonstrated that the elastic recovery rate decreases as the number of step increases. The reason is that with an increasing number of steps, larger energy loss to deform polymer and will result in small elastic recovery. Figure 5 clearly shows that the multi-step imprinting has a significant impact on the elastic recovery.

Table 4. Effect of the number of printing steps on recovery deformation of micro-structure

No.	Multi- step ratio	Width/(mm)	Depth/(mm)	Area after the rebound	Area before the rebound	φ %
1	0.5	1.003010	0.471099	0.742280	0.785398	5.60
2	0.3:0.2	1.012480	0.479123	0.761995	0.785398	2.98
3	0.15 :0.15:0.15	0.972328	0.503947	0.769693	0.785398	2.00

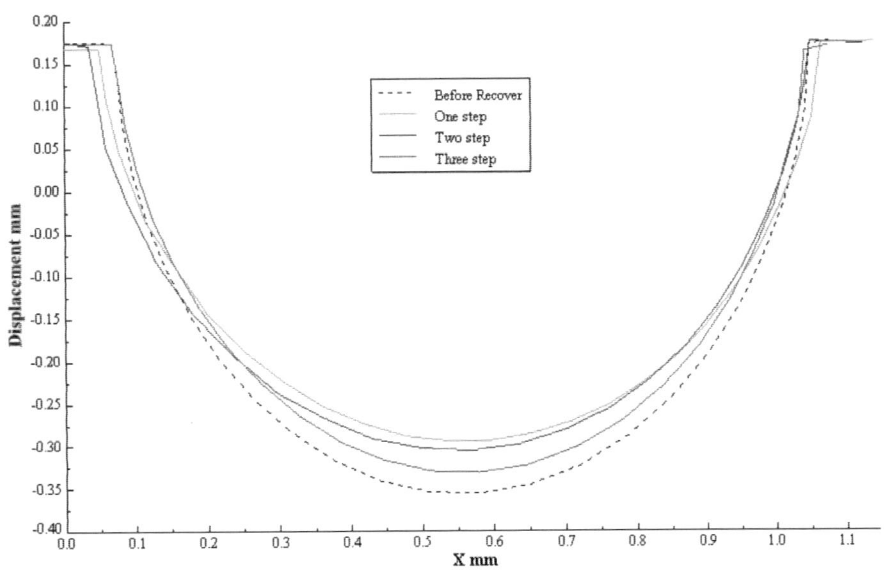

Figure 5. Effect of the number of printing steps on recovery deformation of micro-structure

4.2 Effect of the imprint depth ratio in the two-steps imprint process

The effect of the imprint depth ratio on recovery deformation of micro-structure, which the ratio is defined as the ratio of the imprint depth in the first step to its amount in the second step was studied and the results showed a remarkable effect. Figure 6 shows the change of the imprint depths versus time for different two-steps process, which illustrated the displacement change at the selected point A (shown in figure 3) with different imprint ratio from the beginning to the demold.

Table 5. Effect of the imprint depth ratio on recovery deformation of micro-structure

No.	Two step ratio	Width/(mm)	Depth/(mm)	Area after the rebound	Area before the rebound	φ %
1	0.3:0.2	1.01248	0.479123	0.761995	0.785398	2.98
2	0.2:0.3	1.01209	0.469291	0.746073		5.01
3	0.35:0.15	1.00984	0.468767	0.743583		5.32
4	0.15:0.35	1.00444	0.464366	0.732663		6.71

The detailed elastic recovery rate (φ) with different imprint depth ratio is tabulated in table 5. It is found that the elastic recovery rate (φ) is 2.98 % at imprint depth ratio 0.3: 0.2 (0.3mm in depth at the first step and 0.2 mm at the second step) which is the least value achieved. The elastic recovery rate (φ) is 5.01 % if the imprint depth ratio is reversal with 0.2: 0.3. However, this rule is also applied to the imprint depth ratio of 0.35: 0.15, when the imprint depth at the first step is larger than that at the second step, will result in small elastic recovery.

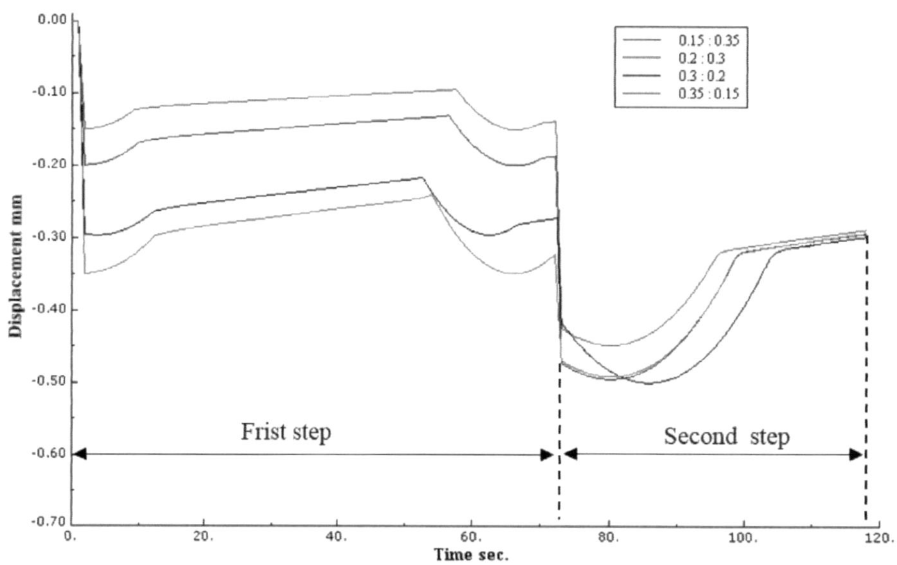

Figure 6. The imprint depths during the two-steps imprint process with time

In addition, the imprint depth ratio for two-step process has a profound effect on the elastic recovery, as shown in figure 7. By using different imprint depth ratio through steps, the elastic recovery rate will be different. These imply that the imprint depth ratio and their order may have not negligible effect on elastic recovery.

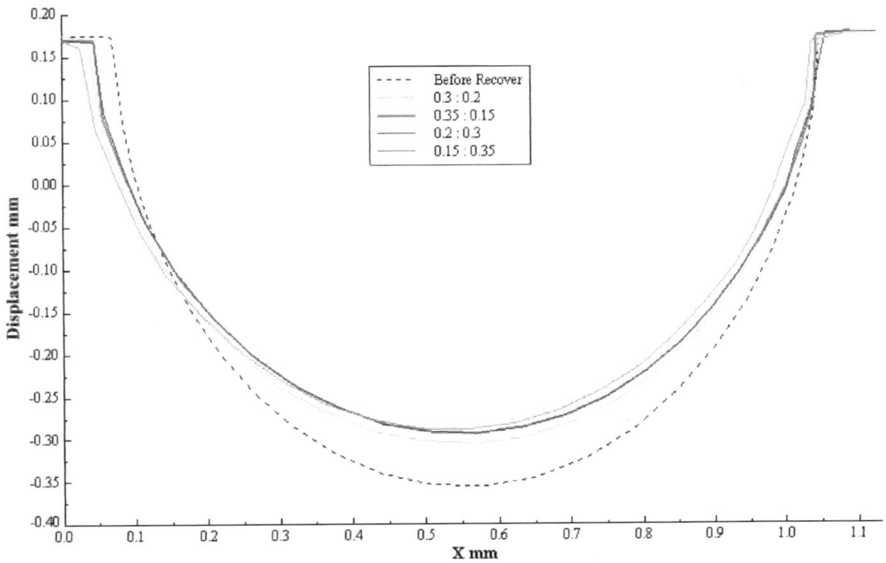

Figure 7. Effect of the imprint depth ratio on recovery deformation of micro-structure

4.3 Effect of the hold-on time on recovery deformation of micro-structure

The molding quality of the fabricated part is often influenced during the pressurization and demolding stages. To enable the polymer forming better in the hot press forming process, the requirements of the fine structure cavity and the efficient mold release are needed in general. Longer contact time between the rolling mold and polymer substrate provoked sufficient heat transfer before a demolding, which most likely enhanced the polymer cooling and will result in a decrease recovery deformation of micro-structure [16-17]. Table 6 shows the effect of hold-on time on the recovery deformation of micro-structure. It is observed that the value of the recovery deformation decreases as hold-on time increases as the hold-on time is less than 50 seconds. However, it is found that the value of the recovery deformation increases to 1.42% at 60 seconds of hold-on time and then the value of recovery deformation hold is kept as a steady value, as shown in figure 8.

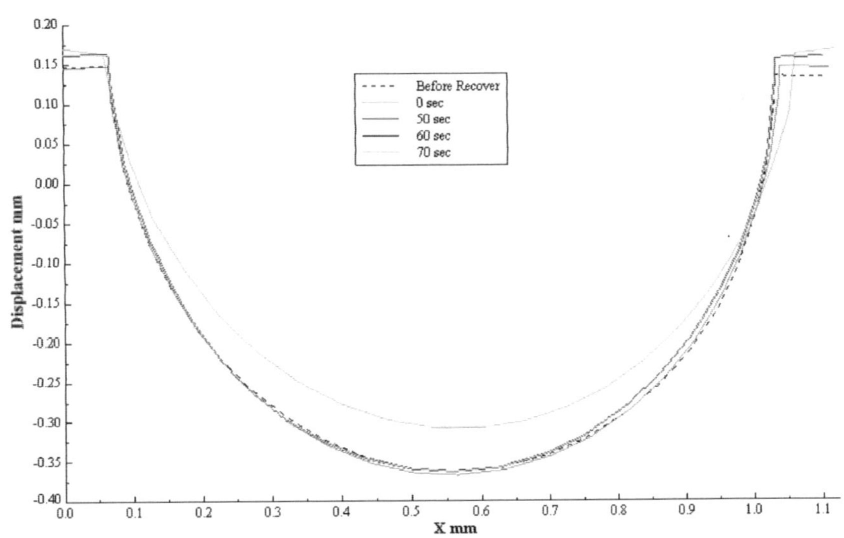

Figure 8. Effect of the hold-on time on recovery deformation of micro-structure

Table 6. Effect of the hold-on time on recovery deformation of micro-structure

No.	Hold-on time/（sec）	Width/(mm)	Depth/(mm)	Area after the rebound	Area before the rebound	φ %
1	0	1.00301	0.471099	0.74228	0.785398	5.60
2	50	0.974171	0.513831	0.786397	0.785398	0. 18
3	60	0.966131	0.524901	0.798587	0.785398	1.42
4	70	0.968133	0.524901	0.798238	0.785398	1.63

5. Conclusion

In the paper, the genetic algorithm (GA) was employed to obtain a Prony series of the viscoelastic generalized Maxwell model in terms of experimental data of TOPAS®COC material. Using commercial nonlinear finite element software package of ABAQUS, a rolling hot forming model with different processes were established and simulated. Several conclusions can be made as follows: The effect of a multistep imprint on elastic deformation recovery is remarkable. To reduce elastic deformation recovery, the multistep processes are recommended. The imprint depth ratio for the two-steps imprint process and their sequencing have significant effects on elastic recovery. The mold quality of the fabricated part will be affected by hold-on time as well.

6.References

[1] Chou S, Krauss P and Renstrom PJ 1996 *J. Vac. Sci. Technol* B 14(6), p 4129–33.
[2] Hirai Y, Yoshida S and Takagi N 2003 *J. Vac. Sci. Technol* B 21, 2765 10.1116/1.1629289.
[3] Song Z C et al 2008 *Microsystem Technologies-Micro-and Nanosystems-Information Storage and Processing Systems* 14(9-11): p. 1593-7.
[4] Young W 2005 *Microelectron. Eng.* 77 p 405–411.
[5] Sirotkin V et al 2006 *Microelectronic Engineering* 83 p 880–3.
[6] Rowland H, Sun C, Schunk R and King P 2005 *Micromech. Microeng.* 15 p2414–25.
[7] Scheer C, Bogdanski N, Wissen M, and Hirai Y 2006 *Microelectronic Eng.* 83 p843–6.
[8] Liu C, Li J, Liu J and D L 2010 *Wang Microelectronic Engineering* 87 p200–207.
[9] Narijauskaitė B, Palevičius A, Gaidys R, Janušas G and Šakalys 2013 *Sensors* 13, 11229- 42.
[10] Jena R, Chen X, Yue C Y and Lam Y C 2011 *Micromech. Microeng.* 21 085029 (12pp).
[11] Hongbo L 2018 *Intech Open*, http://dx.doi.org/10.5772/intechopen.72860.
[12] Yang, S Yu, S Cho and Maenghyo 2009 *Japanese Journal of Applied Physics*. 48. 10.1143/JJAP.48.06FH03.
[13] Qu F, Jiang Z and Lu H 2015 *Journal of Applied Mathematics Volume*, Article ID 424131, 7 pages http://dx.doi.org/10.1155/2015/424131.
[14] Cheng J F and Chen X 2017 *International Journal of Materials Science and Applications* pp. 207-11. doi: 10.11648/j.ijmsa.20170604.17
[15] Cheng J F, Chen X, Deng Y and Li D 2018 *China Plastic* CN11-1846/TQ ISSN 1001-9278.
[16] Yeo, L P et al 2009 *Microsystem Technologies-Micro-and Nanosystems-Information Storage and Processing Systems* 15(4): p. 581-93.
[17] Shan X C, Liu Y C, Lu H J, Wang Z F and Lam Y C 2007 *DTIP* p 88-91. ffhal-00257667f (Stresa, lago Maggiore, Italy).

IOP Conf. Series: Materials Science and Engineering **813** (2020) 012006 doi:10.1088/1757-899X/813/1/012006

Safety Assessment of Steam Generator in Hydrogen Production Unit under Local Overtemperature Conditions

DU Liang[1, a], YAN He[1,b] , ZHAO Wenjing[1,c] , LIU Deyu[1,d] , CHEN Yanze[1,e]

[1] China Special Equipment Inspection and Research Institute, Beijing, China

[a]DULIANG.SME@163.COM, [b]Peteryan76@126.com, [c] 18600298568@163.com, [d] 13911311796@139.com, [e] 750311949@qq.com

Abstract. A steam generator at the bottom of a radiant chamber of a hydrogen production unit is used to heat the conversion gas of the reformer to the boiler feed water, thus producing medium pressure steam. During the inspection process, the color of the temperature indicating coating was found to be abnormal near the manhole block of the equipment. The maximum temperature of the wall of the device was up to 472.7°C by infrared thermal imager. In this paper, the reason of the overtemperature of the steam generator is analyzed. Secondly, the stress strength evaluation based on the conventional strength checking formula and the finite element numerical simulation analysis method is carried out. Finally, the monitoring suggestion and the follow-up test proposal under the high temperature operating condition are given.

1. Introduction

Steam generator is a mechanical device that uses heat energy from fuel or other sources of energy to heat water to produce steam[1-3]. The unreasonable lining structure and material will cause the lining failure, which will cause the high pressure and high temperature hydrogen-containing converting gas to enter the wall from the lining crack, and form a circulating path between the wall and the lining, which will cause the wall overheating and reduce the use intensity. Local overheating is also a common failure form of this kind of steam generator[4-6].Aiming at the over-temperature condition of a steam generator at the bottom of a converter, this paper studies the causes of local over-temperature. It is proposed that the material service strength under the over-temperature condition of the converter should be checked by using the conventional mechanical evaluation based on empirical formulas and the stress intensity checking based on finite element analysis, respectively. Meanwhile, the monitoring measures and inspection ideas under the reasonable high temperature operation condition are given.

2. Survey of the Steam Generator

The schematic diagram of the steam generator (No.1101-E-102) at the bottom of the radiation chamber of the reformer of a hydrogen plant is shown in Figure 1. The process parameters of the steam generator are shown in Table 1. During inspection, abnormal color of the temperature-indicating coating near the short section of the manhole seat 2 is found. As shown in Figure. 1, the maximum temperature of the local wall of the steam generator reaches 472.7°C, and the maximum temperature is located at Ar1 of Figure. 2.

Preliminary analysis of the reasons for the over-temperature of this location is as follows: The overtemperature is located at the short section of the manhole seat. The inner seal of the part is completely insulated by the manhole refractory brick. Because the curvature of the manhole brick is different from that of the manhole brick, the gap between the manhole brick and the lining is easy to occur. The transformed gas enters into the weld joint between the manhole and the shell, resulting in local overtemperature of the position. Because it is impossible to stop and repair immediately, the strength of the overtemperature part should be calibrated for safety reasons. The stress level of the structure at the over-temperature position was determined and the safety evaluation was carried out.

Table 1. Basic data table

Name	Converter Steam Generator	
/	Shell side	Tube side
Maximum working pressure (MPa)	4.4	2.5
Operating temperature (°C)	256	300
Design Temperature (°C)	263	300
Design Pressure (MPa)	4.9	3.77
Corrosion margin (mm)	3	3
Medium	Boiler feed water, steam	Converted gas

Figure 1. Overtemperature position

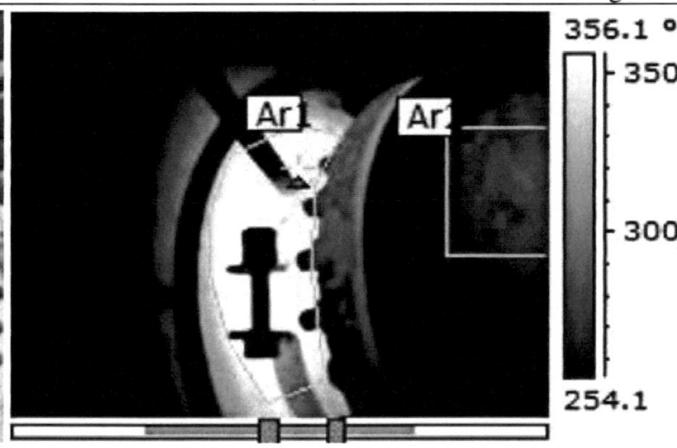

Figure 2. Infrared thermal imaging test results

3. Stress Intensity Calculating and Checking in Conventional Way

According to the theory of elastic thin shell, the connection structure between the manhole seat and the inlet box of steam generator belongs to the cylindrical shell hole reinforcement structure. Standard JB4732-1995（R2005） "Steel Pressure Vessel-Analysis and Design"[7] gives the stress analysis mechanical model of this kind of structure, as shown in Figure. 3, which is applicable to cylindrical shells with radial smooth nozzles under internal pressure.

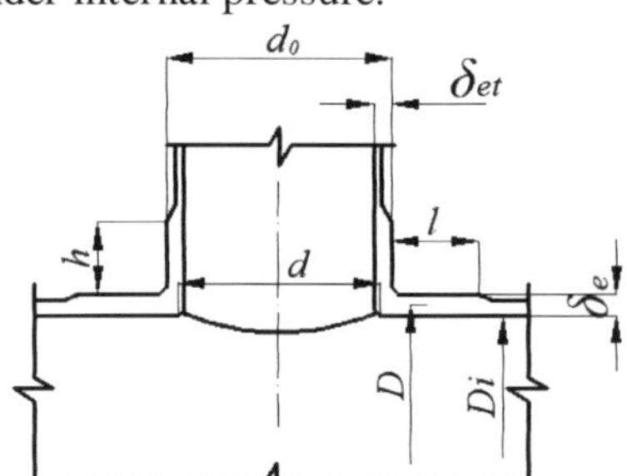

Figure 3. Mechanical model of cylindrical shell with hole reinforcement structure

Dimensions and related representations in the figure are as follows:

D_i ——Internal diameter of cylindrical shell,take as 1600mm.

d_0 ——Outer diameter of nozzle,take as 610mm.

K ——Strength concentration factor of primary and secondary stresses.

K_m ——Intensity concentration factor of primary local membrane stress.
P——Operation pressure, take as 2.5Mpa.
P_d ——Design pressure, take as 3.77MPa.
D——Diameter of the medium surface of cylindrical shells,mm.
d ——Diameter of middle surface of nozzle,mm.
S_m ——Design stress strength of materials,Mpa. The allowable stress of 15CrMo (plate thickness < 300mm) material at 470°C is 110 MPa (yield strength of 15CrMo at 470°C is 165 MPa, safety factor is 1.5).
δ_e ——Effective thickness of cylindrical shells,take as 37mm.
δ_{et} ——Effective thickness of nozzle,take as 62mm.
ρ_0 ——Opening rate, take as $\rho_0 = d / D = 0.33$.
λ ——Parameter, $\lambda = \rho_0 \sqrt{D / \delta_e} = d / \sqrt{D\delta_e}$.

In JB 4732, the process of checking the stress intensity of cylindrical shell reinforced structures with holes based on elastic thin shell theory is as follows:

a) Calculate D, d :

$$D = D_i + \delta_e = 1600+37=1637\text{mm};\ d = d_0 - \delta_{et} = 610-62=548\text{mm}$$

b) Calculate ρ_0、λ、δ_{et} / δ_e :

$$\rho_0 = d / D = 0.33,\quad \lambda = \rho_0 \sqrt{D / \delta_e} = d / \sqrt{D\delta_e} = 2.23,\quad \delta_{et} / \delta_e = 62/37 = 1.68$$

Inquire K_m and K based on ρ_0、λ、δ_{et} / δ from curve J-2 set given in Ref.6, $K_m = 2.25$、$K = 3.0$.

c) Calculate:

Primary general membrane stress in conservative way: $S_{\text{I}} = \dfrac{P_d D}{2\delta_e} = 83.4\text{MPa}$

Primary local membrane stress: $S_{\text{II}} = K_m \dfrac{PD}{2\delta_e} = 124.4\text{MPa}$

Stress strength of primary plus secondary: $S_{\text{IV}} = K \dfrac{PD}{2\delta_e} = 165.9\text{MPa}$

d) Stress check:

$$S_{\text{I}} \le S_m = 110\text{MPa},\ S_{\text{II}} \le 2.2S_m = 242\text{MPa},\ S_{\text{IV}} \le 2.6S_m = 286\text{MPa}$$

From the above analysis, it can be seen that the material strength at the local over-temperature of the short segment of the manhole seat can meet the permissible requirements under the condition of over-temperature 470°C obtained by empirical formula.

4. Stress Intensity Calculating and Checking Based on ANSYS

In order to further clarify the distribution of stress field in the over-temperature part of the steam generator tube box and obtain more accurate stress intensity value in the over-temperature part, this paper based on the finite element analysis software ANSYS to carry out detailed stress analysis of the inlet section of the steam generator tube box, and establish the finite element model.

The finite element model element selected in this analysis is SHELL181 two-dimensional planar quadrilateral element. This element is suitable for thin plate and shell structure simulation with a certain thickness. It can output the film stress, bending stress and peak stress of the specified element. It overcomes the shortage of linearization of the thickness direction stress of the general plate and shell element, and can better simulate the real stress of the tube-box structure, and directly give the stress value needed for strength checking.

Considering the location and the mechanical boundary conditions in JB4732, the analysis and selection of loads mainly include internal pressure under operating conditions, vessel weight, additional loads (gravity loads applied in the form of equivalent density for insulation materials, linings,

IOP Conf. Series: Materials Science and Engineering **813** (2020) 012006 doi:10.1088/1757-899X/813/1/012006

accessories, etc.) and wind loads. The material performance parameters of the pipe box section are shown in Table 2.

Table 2. Material performance parameter

	Material	Modulus of elasticity E（MPa）	Poisson ratio	Density（Kg/m^3）
#1	15CrMo	1.87×10^5	0.31	7.85×10^3

4.1. Stress Intensity under Operation Condition

Figure. 4 and Figure.5 show the stress intensity (the third strength theory) of the whole structure of the steam generator tube section under the combined action of mechanical load, mechanical load and temperature load under operating conditions, respectively. It shows that the maximum stress intensity of the structure is 234.8 MPa under mechanical loading under operating conditions. The maximum stress intensity of the structure under the combined action of mechanical load and temperature load is 236.1 MPa.

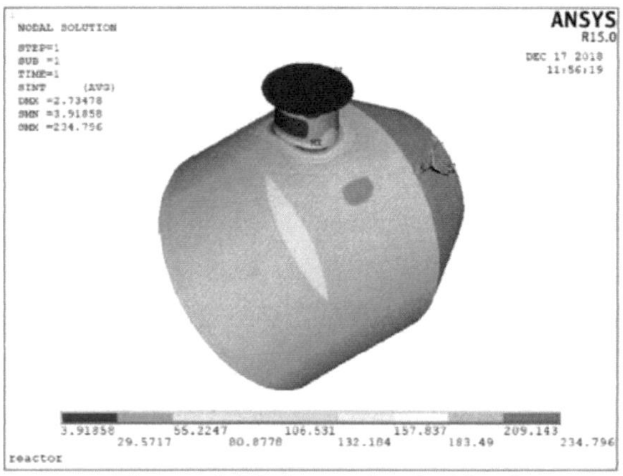

Figure 4. Under mechanical load

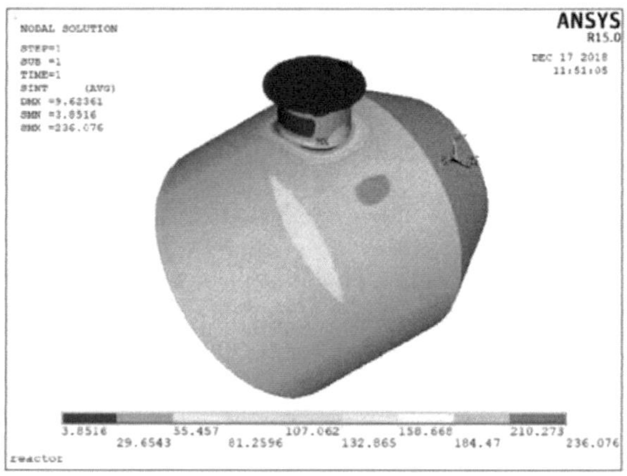

Figure 5. Under mechanical and temperature loads

Figure. 6 and Figure. 7 show the equivalent stress of steam generator tube box under the combined action of mechanical load, mechanical load and temperature load, respectively. The results show that the maximum stress intensity of tube box is located on the tube body.

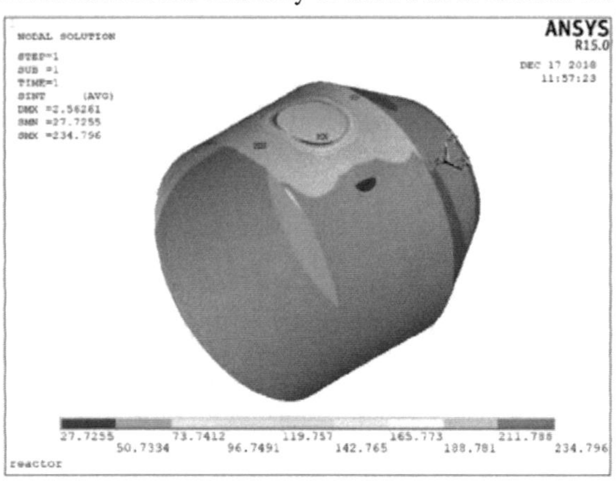

Figure 6. Under mechanical load

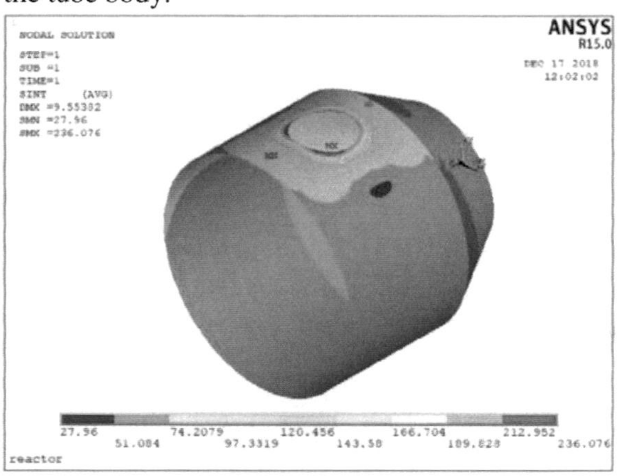

Figure 7. Under mechanical and temperature loads

Figure. 8 and Figure. 9 show the equivalent stress at the manhole joint of the steam generator tube box under the combined action of mechanical load, mechanical load and temperature load, respectively. The results show that the stress intensity of the manhole joint under the combined action of mechanical load and temperature load is lower than that under the action of mechanical load alone. Temperature load decreases the local stress intensity at the short section of manhole. In this paper, the structural stress level of the short section of manhole, i. e. the over-temperature part, is evaluated.

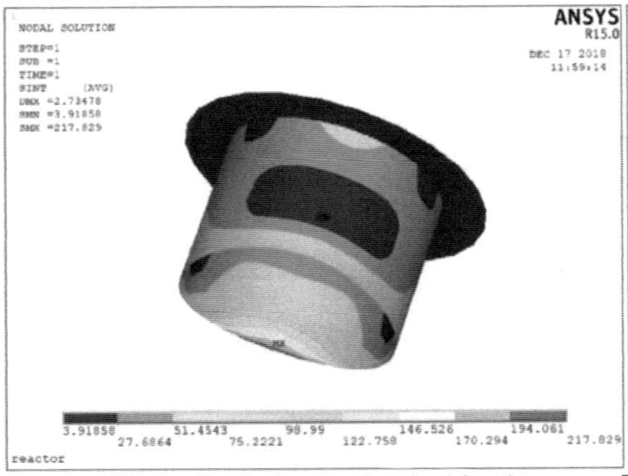

Figure 8. Under mechanical load

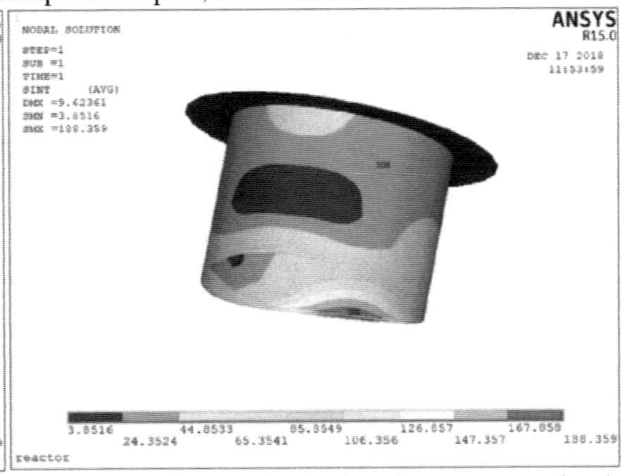

Figure 9. Under mechanical and temperature loads.

4.2. Intensity Check of Primary Local Membrane stress

The analysis results of Figure. 10 show S_{II} is 210.38 MPa, which $S_{II} \le 2.2S_m$=242MPa ,where the allowable stress S_m of the material corresponds takes as 110 MPa at 470 °C.

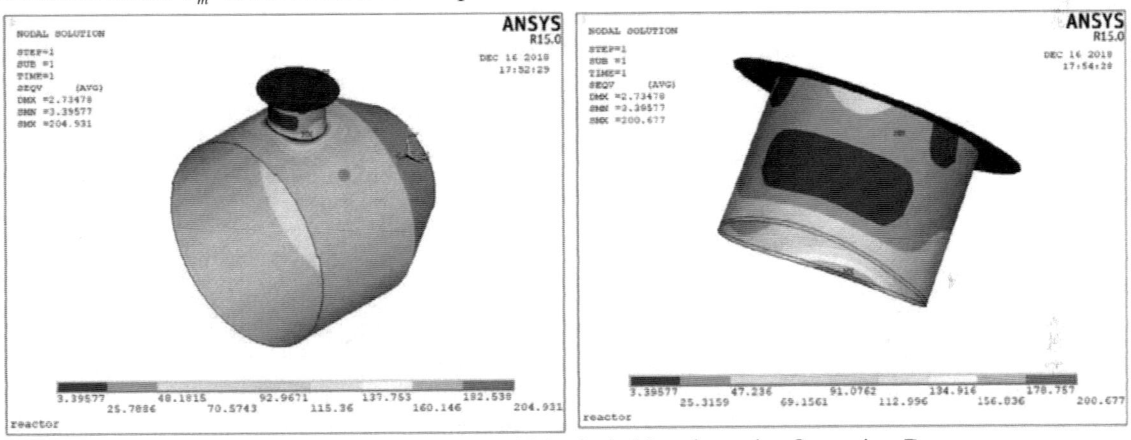

Figure 10. Stress Diagram of Manhole Nozzle under Operating Pressure

4.3. Primary and Secondary Stress Intensity Checking

According to the standard requirements, the primary and secondary stresses caused by the combined action of pressure, temperature and external load of nozzle diameter are classified as S_{IV}, which takes as 193.96MPa, so that $S_{IV} \le 2.6S_m$=286MPa which compliance with standard requirements.

5. Treatment of Local Thermal Stress

For the local thermal stress, the structural deformation will not be obvious after the structural restraint is removed. The stress only needs to be considered in the fatigue analysis, that is, the local thermal stress belongs to the peak stress. The stress in a small area of local superheat on the vessel wall is a typical local thermal stress.

For nozzles with stiffening rings and non-integral structures, fatigue analysis can be avoided if the total number of the following cycles does not exceed 400 when the tensile strength R_m of steel at room temperature is less than 550 MPa: a) Predict (design) the number of pressure cycles during the whole process, including start-up and parking. b) The number of times the pressure fluctuation exceeds the expected number of cycles of working pressure of 15% of the designed pressure. c) The effective number of temperature fluctuations (170°C) between any two adjacent points, including the takeover; multiplied by the coefficient of 8.0.

According to the process parameters, the total number of cycles above is less than 400, so the equipment can avoid fatigue analysis.

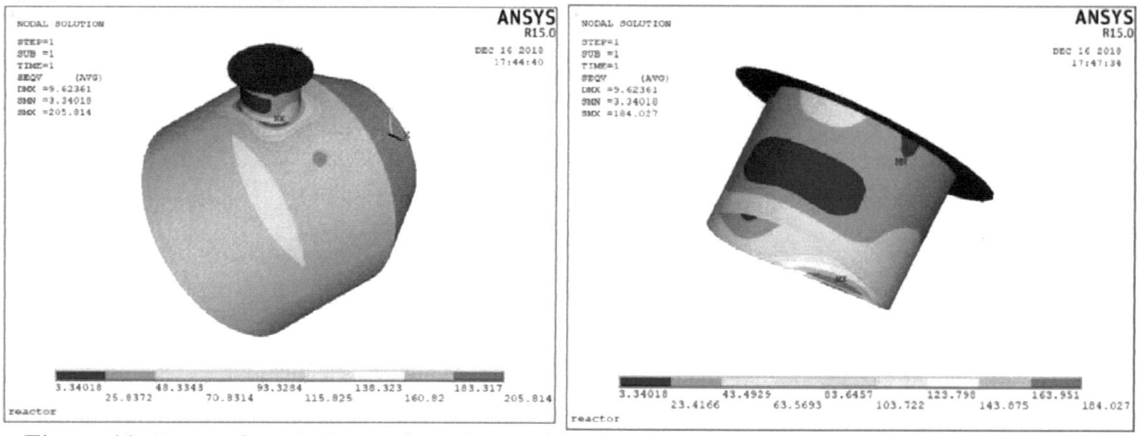

Figure 11. Stress of manhole nozzle under combined action of operating pressure and temperature

6. Conclusion

Through the calculation and analysis of the converter steam generator (No. 1101-E-102) based on the correlation strength checking formula and the finite element numerical method, it is shown that the primary and secondary stress strength of the pressure vessel meets the relevant standard requirements. Although the local structure is overheated (470°C), it is still lower than the maximum service temperature of 15CrMo material 550 °C, which causes the material allowable stress to decrease to a certain extent, but the temperature is still lower than the upper limit of the service temperature of 15CrMoR steel given in GB 150[8]. After checking, the stress intensity of the overheated part still meets the allowable stress requirements of the container under operating conditions.

7. References

[1] LU Chao: Reason Analysis and Treatment of Internal Leakage in Steam Generator of Converter Gas in Hydrogen Production Unit, edited by Petroleum Refinery Engineering(2019),in press.

[2] WANG Shude, ZHANG Shaoliang and YE Xiaodong: Problems and Analysis of Converter Steam Generator in Hydrogen Plant, edited by Petroleum Processing and Petrochemicals (2005),inpress.

[3] LIU Yijia: Modeling and Simulation of steam generator, edited by Oil and Gas Field Surface Engineering (2016),in press.

[4] LI Hongyan, LI Guangwei: Detection of Local Heat Treatment Deformation of Supporting Plate on the Top of Steam Generator, edited by Pressure Vessel Technology(2018),in press.

[5] SUN Guohang, ZHAO Jiuguo, YANG Xinbo,etc: Study on Material Performance of Waste Heat Boiler after Overtemperature Operation, edited by Selection of Advanced Technology for Pressure Vessels at the Sixth National Pressure Vessel Academic Conference.(2005).in press.

[6] QIN Xiaodong: Study on Overtemperature of Superheater of High Temperature and High Pressure Boiler with Pure Fuel Blast Furnace Gas, Shanghai Jiao Tong University,(2008).

[7] JB4732-1995（R2005）,Steel Pressure Vessel-Analysis and Design.

[8] GB150-2011: Pressure Vessel.

IOP Conf. Series: Materials Science and Engineering **813** (2020) 012007 doi:10.1088/1757-899X/813/1/012007

Distributed Electric System Control Strategy Optimization

Chunping Chen [1,2], Jie Wang[2] and Jie Zhang [1,a]

[1]Air and Missile Denfense College, Air Force Engineering University, Xi'an, China
[2]Shanghai Institute of Precision Metrology and Testing, Shanghai, China

[a]afeu_zhangjie@163.com

Abstract. With the increasing energy crisis and environmental pollution, this paper concentrates on the energy management and power control of DC microgrid due to its advantage in the utilization of renewable energy resources (RES). However, the traditional centralized control relies heavily on communication since the distribution of most RES is scattered, and the communication failures may cause serious control error, therefore, a distributed and cooperative control strategy without communication is proposed for DC microgrid in this paper. The proposed strategy not only maintains the bus voltage and power balance in a cooperative way through the regulation of each micro-sources, but also solves theunbalance problem between ESUs, meanwhile, this method is suitable for DC microgrid with multiple PVs and ESUs since the communication is not necessary, which is benefit for the plug and in of micro-sources. Finally, some simulation models are built on Matlab/Simulink platform, and the performance of suggested solution is verified through a set of simulation experiments, the result turns out to be good and confirms our expectation.

1. Introduction

With the continuous development of the national economy, the traditional large power grid has been rapidly developed as a major power supply channel for a long period of time. However, in response to the increasing demand for electricity, with the increasing scale of the power grid, the cost of centralized large-scale power grids has gradually increased, the operation difficulty has increased, and the control system has become more complicated, so that the reliability of power consumption of diverse users has been challenged [1-8]. In particular, some large-scale power outages that have occurred in recent years have exposed the shortcomings of large power grids in terms of safety and reliability, such as the large-scale long-term power outage caused by the southern ice disaster in 2008, 7.30 in India in 2012. Large power outages, etc., not only caused economic losses, but also affected social stability. In addition, with the continuous advancement of industrial production, traditional non-renewable energy sources such as coal, oil and natural gas have been over-exploited, and with the increasing demand for load, human beings will inevitably face the crisis of energy shortage in the long run. At the same time, traditional fossils Excessive use of fuel not only causes irreparable pollution to the natural environment, changes the global climate, but also damages human health and endangers human survival. Therefore, renewable new energy sources with abundant reserves, clean and pollution-free advantages, such as solar energy, wind energy, and ocean energy, have received extensive attention, and have become the development trend of future energy utilization. Compared with traditional firepower and hydropower generation, renewable energy such as solar energy and wind energy has the characteristics of small environmental pollution, flexibility and never exhaustion. Therefore, improving the penetration rate of renewable energy in the power grid is one of the important signs of power grid progress [9].

In view of the advantages of microgrid in the use of new energy, some researchers believe that microgrid will be an important part of the next generation of smart grid [10]. At present, the microgrid structure studied at home and abroad includes AC microgrid, DC microgrid and AC/DC hybrid microgrid. Since the traditional large grid mainly exists in the form of communication, the AC microgrid is still the main form of microgrid. Research on the exchange of microgrids has also received extensive attention.

Compared with AC microgrid, DC microgrid has many advantages, which can be summarized as follows [9]:

(1) The DC power supply in the DC microgrid can be connected to the DC bus through a DC/DC converter, and only one DC/AC converter is needed for energy exchange between the DC microgrid and the large grid, effectively reducing the converter [11]. The number of them reduces the difficulty and cost of microgrid control;

(2) Since the DC microgrid [12] does not need to consider frequency, phase, reactive power and other issues with respect to the AC microgrid, it mainly involves the control of active power, so the bus voltage becomes the unique identifier of the system power balance in the DC microgrid [13]. As long as the bus voltage is stabilized, the power balance of each unit in the DC microgrid system [14] can be realized, and the control is simple and reliable, and the power flow and energy scheduling can be realized quickly.

(3) DC conversion can significantly reduce the distortion and clutter that often occur in AC/DC conversion process, so DC microgrid can significantly improve the quality of power supply. In addition, the presence of the grid-connected inverter effectively separates the DC microgrid [15] from the large grid, so that the DC microgrid can be used as an independent whole, which can quickly disengage from the faulty large grid and operate on an island, and implement self-sufficiency. And can be re-integrated into the large grid in time to operate together, which improves the reliability of the power supply and the flexibility of the system.

(4) DC microgrid has the advantages of simple design, low cost and high efficiency in practical engineering. Under the traditional large power grid pattern of AC transmission, the DC microgrid [16] shows good adaptability on the micro-power source that uses distributed power generation, especially in the form of DC output and avoids AC transmission by means of DC transmission [17]. DC microgrid has a very broad prospect in the development of new energy and smart grid. At present, the research on the basic theory and practical technology of DC microgrid is still in the preliminary stage, and the actual application has not yet begun. In order to give full play to the advantages of DC microgrid, it is necessary to carry out more in-depth research.

2. Related Research

DC microgrid has attracted extensive attention from researchers because of its own advantages. Research on DC microgrid has become one of the hot research areas of power systems in recent years, including many in the US, Japan, Europe, and China. A large amount of research has been invested in countries and regions. Scholars from various countries have put forward a lot of ideas on the basic theory of DC microgrid and the basic theory of micro-power supply. At the same time, they have made many contributions in the research direction of the structure and operation mode of DC microgrid, and initially established various distributed Energy, micro-grid mathematical models and corresponding simulation analysis tools, and different control methods for different DC microgrid models. In addition, the researchers conducted a large number of experimental tests and built a number of experimental demonstration projects, such as the US Mandrecht microgrid and the Zhejiang Zhoushan Polar Island microgrid project to verify the correctness of the theoretical research and technical aspects and The feasibility of the economic level makes the DC microgrid more practical and has made a huge breakthrough.

At present, research on DC microgrid at home and abroad mainly focuses on model establishment and simulation of microgrid, control strategy of microgrid, energy management and reliability evaluation of microgrid. This paper takes DC microgrid with multiple photovoltaic cells, wind power generation and energy storage system as the research object, first analyzes the principle and operation characteristics of these micro-sources, and then studies the DC micro-grid in networking and island operation without communication. The distributed coordinated control in the mode ensures that the power balance between the micro power supply, the load and the energy storage system in the DC microgrid [18] can be stabilized, the bus voltage can be stabilized at the rated value, and the energy storage unit is passed through the energy storage system. The power distribution realizes the energy storage balance between the energy storage units. The specific research content can be divided into the following three parts:

(1) By analyzing the working principle of two kinds of micro-power sources, such as photovoltaic cells, batteries and their respective power electronic interface devices, the mathematical model is established, and the simulation model is built based on Matlab/Simulink platform to study output. The characteristics lay the foundation for subsequent controller research and overall strategy research on microgrid power balance.

(2) Studying how multiple photovoltaic power sources and multiple energy storage units in a DC microgrid coordinately control the DC bus voltage. In the absence of a communication connection, each micro power supply is based on the different states of the DC bus voltage and their respective local information. Adjust its control strategy to achieve voltage distributed control of the system during network and island operation and smooth switching between various working modes.

(3) Based on the maintenance of the bus voltage at the rated value, the coordinated control strategy of the energy storage system based on local information is studied. By properly distributing the power, the energy storage balance between the energy storage units is realized, and some energy storage in the system is prevented. The unit is in an over- or over-charge condition.

The development trend of renewable energy sources such as photovoltaics and wind power in the future energy structure and the positive role played by microgrids in the use of distributed energy are introduced. The advantages and disadvantages of DC and AC microgrids are described. After comparison, the DC microgrid [19] composed of photovoltaic and energy storage is determined as the research scope of this paper. In addition, the domestic and international research status of DC microgrid is introduced, which involves microgrid modeling and control technology. Finally, the research content of this paper is briefly introduced.

3. Photovoltaic Power Generation System Control Strategy

According to different operating modes of the DC microgrid, the photovoltaic power generation system can be divided into maximum power point tracking (MPPT) and power reduction (off-MPPT) control. In order to lift off the energy utilization rate of the photovoltaic cell, when the output power of the system is less than or equal to the power required for the equivalent load, such as the photovoltaic start-up phase, the illumination blessing is weak, and the load is large, that is, the . At the time, the light supply system operates in the MPPT mode; when the photovoltaic power generation system is redundant in power generation. That is, when the photovoltaic operation is in the power reduction control mode, and the time load and the energy storage unit are both "charge" nodes. For two different control requirements of photovoltaic systems, corresponding control strategies should be designed to ensure stable operation under different working conditions.

4. MPPT Control of Photovoltaic Power Generation System

In addition to its internal characteristics, the energy efficiency of photovoltaic cells is also affected by factors such as light amplitude, temperature and load. According to the PV output characteristics shown in Figures 2.4 to 2.7, the output power of photovoltaic cells is different. Under the conditions of illumination amplitude and humidity, different nonlinear characteristics are presented. Under different external conditions, the photovoltaic cell can operate at different and unique maximum power points, and the output curve is a single peak function with the maximum power point as the extreme value. According to the circuit principle, when the output impedance of the photovoltaic cell and the load impedance are equal, the photochemical battery obtains the maximum output power. It can be seen that the actinic MPPT process is actually the process of matching the output impedance and the load impedance. Since the output impedance is affected by environmental factors, the load impedance needs

IOP Conf. Series: Materials Science and Engineering **813** (2020) 012007 doi:10.1088/1757-899X/813/1/012007

to be adjusted in real time through the control method to track the output impedance and realize MPPT control. The MPPT control for actinic is achieved by controlling the duty cycle of the Boost converter and adjusting the equivalent load impedance.

5. Several common MPPT methods

According to the intelligence and complexity of the control method, it can be roughly divided into three categories.

5.1. Open-loop MPPT Method Based on Output Characteristic Curve

The constant voltage tracking method and the short circuit current proportional coefficient method are two commonly used open-loop MPPT methods. When the humidity is constant, even if the illumination amplitude changes, the open circuit voltage of the photovoltaic cell is almost unchanged. Based on this characteristic, the constant voltage tracking method is based on the maximum power point voltage of the actinic battery. The approximate linear relationship between is approximated. When the illumination amplitude is constant, the short circuit current of the photovoltaic cell is basically unchanged with the change of temperature. Based on this characteristic, the short circuit current proportional coefficient law is approximated according to the approximate linear relationship between the maximum power point current and the current. The two methods are simple and easy to implement, but because they are based on an approximate linear relationship, the photovoltaic cell does not work at the true maximum power point, and the load needs to be disconnected during the measurement, which causes power loss, and the measurement is complicated. Class methods are not practical.

5.2 Self-optimizing MPPT method

Disturbance observation method and conductance increment method are commonly used self-optimization methods for MPPT. Based on the $P-U$ unimodality of the curve, there is a maximum power point, $dP_{pv} / dU_{pv} = 0$ and dP_{pv} / dU_{pv} the symbols at the maximum power point are different. By disturbing the output voltage (or current) of the photochemical cell, the disturbance observation method and the conductance increment method respectively observe the photovoltaic cell output. The sign of the change ΔP_{pv} in power, ΔP_{pv} and dP_{pv} / dU_{pv} then according to the sign of and the sum to determine the direction and step size of the next clever (or current) disturbance, and then repeat, and finally make the battery work in the MPPT state. In essence, the difference between the work and the voltage of the photovoltaic cell output voltage $dP_{pv} / dU_{pv} = 0$ is found, and the working point U_{pv} (or I_{pv}) is found. The main difference between the two is the calculation of the power difference. These two methods are simple and have few parameters ΔP_{pv}, which are independent of the characteristics and parameters of the photovoltaic cell. When the external environment changes, the system can track its changes smoothly, but the tracking accuracy and speed are affected by the initial voltage value and the disturbance step.

5.3 Intelligent MPPT Method

The intelligent control methods such as fuzzy control and neural network are applied to the MPPT control of the unknown and nonlinear photovoltaic system. The system can respond quickly to changes in the external environment and obtain good robustness and adaptability. However, the performance of fuzzy control is very dependent on the selection of the rules, which is greatly influenced by the designer's experience [20]. The intelligent algorithms such as neural networks need to process a large number of data samples, which boils down to the optimal control path, and the learning time is longer. It may not achieve the purpose of learning. In general, the good control performance of intelligent algorithms relies on long control cycles, strong hardware facilities and designer experience, so it is difficult to obtain practical applications in modern industrial control.

6. Disturbance Observation Method with Variable Step Size

Based on the above discussion, the MPPT method of photovoltaic power generation system is still dominated by the classical control theory. The control block diagram of the photovoltaic power generation system is shown. The control signal output by the MPPT algorithm and the PV output voltage obtained by feedback are used as double closed loop. The input signal of the voltage outer loop is controlled to adjust the change of the non-station tracking signal. The disturbance observation method has been widely used because of its simple algorithm and easy implementation. However, it is greatly affected by the disturbance step. If the disturbance step is large, it can ensure fast system tracking speed, but it is easy to be near the maximum power point. A certain amplitude of oscillation; while the disturbance step is too small, although it can reduce the amplitude of the oscillation, it will reduce the tracking speed of the system and cannot adapt to the rapidly changing environment. The MPPT algorithm in this paper adopts a variable step size perturbation observation method, and the step size is set to be proportional to , that is, the step size of the disturbance voltage is automatically changed according to the slope of the curve of the photovoltaic cell, so that the work is performed. When the point is far from the maximum power point, the voltage changes in a large step ensure the rapidity of the MPPT. When the working point is close to the maximum power point, the voltage step becomes smaller, and the oscillation near the maximum power point can be suppressed. As shown in Figure 4.2, where indicates the direction of change of the step size, and the photovoltaic output voltage at the next moment is calculated by the following formula:

$$U_{pv}(k+1) = Upv(k) + N * K * \tag{1}$$

7. Conclusion

In this paper, a photovoltaic power generation system model is built on Simulink. Under the premise of stable bus voltage control, changing the illumination and temperature conditions is as shown above. From the change of photovoltaic cell output voltage and power, it can be seen that when using the variable step disturbance observation method. Through the voltage disturbance, the maximum operating point of the photovoltaic cell can be quickly and accurately searched, and the tracking speed and accuracy can be transported.

Acknowledgment

This work was supported by the Young Scientists Fund of the National Natural Science Foundation of China under Grant No. 61703412, the China Postdoctoral Science Foundation under Grant No. 2016M602996.

8. References

[1] Medjroubi W, Müller U P, Scharf M, et al. Open Data in Power Grid Modelling: New Approaches Towards Transparent Grid Models. Energy Reports, vol.3, pp. 14-21, 2017.
[2] Panteli M, Trakas D N, Mancarella P, et al. Boosting the Power Grid Resilience to Extreme Weather Events Using Defensive Islanding. IEEE Transactions on Smart Grid, vol. 7, no. 6, pp. 2913-2922, 2017.
[3] Gang H, Wang J, Chen C, et al. Integration of Preventive and Emergency Responses for Power Grid Resilience Enhancement. IEEE Transactions on Power Systems, no. 99, pp. 1-1, 2017.
[4] Kakigano H, Miura Y, Ise T. Low-Voltage Bipolar-Type DC Microgrid for Super High Quality Distribution. IEEE Transactions on Power Electronics, vol. 25, no. 12, pp. 3066-3075, 2010.
[5] Cadini F, Agliardi G L, Zio E. A modeling and simulation framework for the reliability/availability assessment of a power transmission grid subject to cascading failures under extreme weather conditions. Applied Energy, vol. 185, pp. 267-279, 2017.
[6] Wang K, Wang Y, Hu X, et al. Wireless Big Data Computing in Smart Grid. IEEE Wireless Communications, vol. 24, no. 2, pp. 58-64, 2017.
[7] Xiao R, Xiang Y, Wang L, et al. Power System Reliability Evaluation Incorporating Dynamic Thermal Rating and Network Topology Optimization. IEEE Transactions on Power Systems, no. 99, pp. 1-1, 2018.

[8] Bhandari V , Sun K , Homans F . The profitability of vehicle to grid for system participants - A case study from the Electricity Reliability Council of Texas. Energy, 2018:S036054421830642X..

[9] Liu X, Wang P, Loh P C. A Hybrid AC/DC Microgrid and Its Coordination Control. IEEE Transactions on Smart Grid, vol. 2, no. 2, pp. 278-286, 2011.

[10] Yeliz Yoldaş, Ahmet Önen, Muyeen S M , et al. Enhancing smart grid with microgrids: Challenges and opportunities. Renewable and Sustainable Energy Reviews, vol. 72, pp. 205-214, 2017.

[11] Ma T, Cintuglu M H, Mohammed O A. Control of a Hybrid AC/DC Microgrid Involving Energy Storage and Pulsed Loads. IEEE Transactions on Industry Applications, vol. 53, no. 1, pp. 567-575, 2017.

[12] Malik S M, Xin A, Sun Y, et al. Voltage and frequency control strategies of hybrid AC/DC microgrid: a review. Iet Generation Transmission & Distribution, vol. 11, no. 2, pp. 303-313, 2017.

[13] Xu L, Dong C. Control and Operation of a DC Microgrid With Variable Generation and Energy Storage. IEEE Transactions on Power Delivery, vol. 26, no. 4, pp. 2513-2522, 2011

[14] Tucci M, Riverso S, Ferraritrecate G. Voltage stabilization in DC microgrids: an approach based on line-independent plug-and-play controllers. IEEE Transactions on Control Systems Technology, no. 99, 2017.

[15] Tomislav Dragičević, Guerrero J M, Vasquez J C, et al. Supervisory Control of an Adaptive-Droop Regulated DC Microgrid With Battery Management Capability. IEEE Transactions on Power Electronics, vol. 29, no. 2, pp. 695-706, 2013.

[16] Gu Y, Xin X, Li W, et al. Mode-Adaptive Decentralized Control for Renewable DC Microgrid With Enhanced Reliability and Flexibility. IEEE Transactions on Power Electronics, vol. 29, no. 9, pp. 5072-5080, 2014.

[17] Eghtedarpour N, Farjah E. Power Control and Management in a Hybrid AC/DC Microgrid. IEEE Transactions on Smart Grid, vol. 99, no. 3, pp. 1494-1505, 2014.

[18] Sharma R K, Mishra S. Dynamic Power Management and Control of PV PEM fuel Cell based Standalone AC/DC Microgrid Using Hybrid Energy Storage. IEEE Transactions on Industry Applications, no. 99, pp. 1-1, 2017.

[19] Anand S, Fernandes B G, Guerrero J. Distributed Control to Ensure Proportional Load Sharing and Improve Voltage Regulation in Low-Voltage DC Microgrids. IEEE Transactions on Power Electronics, vol. 28, no. 4, pp. 1900-1913, 2013

[20] Wang B, Sechilariu M, Locment F. Intelligent DC Microgrid With Smart Grid Communications: Control Strategy Consideration and Design. IEEE Transactions on Smart Grid , vol. 3, no. 4, pp. 2148-2156, 2012.

AUTHOR INDEX

Institute of Physics
Dirac House, Temple Back
Bristol BS1 6BE UK

ISSN: 1757-8981
ISBN 978-1-7138-2675-0

CURRAN ASSOCIATES INC.
proceedings.com